世界经典建筑设计丛书

办公建筑

韩国 A&C 出版社 著

北京科学技术出版社

Copyright © A&C Publishing Co.,Ltd.
All rights reserved.
Chinese simplified translation copyright Beijing Science and Technology Publishing Co.,Ltd.

图书在版编目（CIP）数据

办公建筑 / 韩国A&C出版社著；北京科学技术出版社译. — 北京：北京科学技术出版社，2019.1
（世界经典建筑设计丛书）
书名原文：Office Building
ISBN 978-7-5304-9392-2

Ⅰ.①办… Ⅱ.①韩…②北… Ⅲ.①办公建筑－建筑设计－作品集－世界－现代 Ⅳ.①TU243

中国版本图书馆CIP数据核字(2018)第063373号

世界经典建筑设计丛书·办公建筑

作　　者：	韩国建筑世界出版社
策划编辑：	陈　伟
责任编辑：	王　晖
封面设计：	芒　果
责任印制：	张　良
出 版 人：	曾庆宇
出版发行：	北京科学技术出版社
社　　址：	北京西直门南大街16号
邮政编码：	100035
电话传真：	0086-10-66135495（总编室）　0086-10-66113227（发行部）
	0086-10-66161952（发行部传真）
网　　址：	www.bkydw.cn
电子信箱：	bjkj@bjkjpress.com
经　　销：	新华书店
印　　刷：	北京捷迅佳彩印刷有限公司
开　　本：	920mm×1260mm　1/32
字　　数：	237千字
印　　张：	8.5
版　　次：	2019年1月第1版
印　　次：	2019年1月第1次印刷
ISBN 978-7-5304-9392-2/T·969	

定　　价：138.00元

京科版图书，版权所有，侵权必究。
京科版图书，印装差错，负责退换。

>>CONTENTS
办公建筑

KNHC GWANGJU JEONNAM 分公司办公室
TOMOON Engineering & Architects Universe-Top Engineering & Architects
UNSANGDOONG建筑公司_006

KNHC DAEGU GYEONGBUK 分公司办公室
WONYANG Architectural Design Group_020

文化资源中心
WONYANG Architectural Design Group_036

WOONGJIN THINKBIG
Kim in-cheurl ARCHIUM Architect_050

DMC办公楼（TRUTEC）
Barkow Leibinger Architects
CHANGLO Architects_072

数字奇异空间
HEERIM Architects & Planners_084

KORLOY HOLYSTAR 大楼
MANO建筑公司_098

B2Y办公室
Y SPACE_110

JEOLLANAM-DO 政府办公室
G.S.建筑公司 Kim Hyunchul_120

SEOUL 邮政中心办公室（SCPO)
SPACEGROUP HEEERIM Architects & Planners
HANGIL Architects & Engineers
DeStefano Keating Partners有限公司_134

MERITZ 火灾海上保险公司大楼
SHINHAN Architects & Engineers   KEATING/KHANG LLP_146

TAEYOUNG YEOUIDO
大楼
WONDOSHI建筑集团有限公司_154

大韩民国驻华使馆
KUNWON Architects Planners Engineers_164

ENVISION
Design Pure_170

BLOOMING
休息大厅
VOID planning_180

MAC政府大厦
BAUM建筑工程咨询公司　ARTECH Architects
Kim Hong-il_188
HEERIM Architects & Planners_194
SPACE GROUP_200

多功能市政厅
HEERIM Architects & Planners
Kim Jung-gon_204　KIAHAN Architecture_210
KUN-WON Architects　Planners Engineers_216

CHUNGCHEONG-NAM-DO
议会和办公室
AUM & LEE建筑公司_222

海洋生物资源国立学院
HEERIM Architects & Planners　DOSHIN
Architecture_228

GIMPO-YANGCHON
生态中心
YEZU建筑规划公司_236
Idea Image建筑学院_242
NODE建筑规划公司　Lee Han-gi_249

CHEONAN
主题公园建筑群
HAEAHN Architects & Planners
USUN Engineering　DAMOOL Architects_254
AUM & LEE 建筑公司　TOWOO
Architects &Engineers_261
DOSHIIN Architecture　DONGWOO Architects &
Consultants　DS GROUP Architecture_266

\>\> OFFICE

GWANGJU JEONNAM BRANCH OFFICE OF KNHC

Location Chipyeong-dong, Seo-gu, Gwangju, Korea **Site Attribute** Central Commerce, Fire-Prevention, District Unit Plan **Site Area** 4,419.10m² **Building Area** 2,859.26m² **Total Floor Area** 29,701.55m² **Building Coverage Ratio** 64.70% **Floor Area Ratio** 429.41% **Building Scope** B3–15F **Structure** RC, Steel **Exterior Finish** 124 Pair Glass, P.F.W, Aluminium Sheet, Granite, Exposed Concrete **Architecture Design** Choi Ki-chul, Han Nam-soo | TOMOON Engineering & Architects + Oh Kum-yeol | Universe-Top Engineering & Architects + Jang Yoon-gyoo | UNSANGDOONG Architects Cooperation **Design Team** Hong Chang-sung, Lee Seon-jeong, Choi Sang-ik, Jo Dong-ho, Choi Yun-sang **Client** Korea National Housing Corporation **Photograph** TOMOON Engineering & Architects

The office of the Korea National Housing Corporation in Gwangju Jeonnam was expected to play a symbolic role as a connector of the local society for housing complexes, commercial facilities, parks, public facilities and so on. The focus of the plan was expanded to a plan containing openness to the citizens, environment-friendliness, differentiated elevation and active eco-energy system with this symbolism. Light Band expressed the image of the office in Gwangju, which is called as light county, and scenic light emphasized the function of light that was involved in the construction program. The office built in an unique way with its own style has an easy access from tourism spots and the 20m-width road in front of it to be a landmark in that business area. The lower part including a square and an exhibition all was planed to be a space for the local society and external and internal squares were directly connected to each other to support the structure. The representative factor for the connected external and internal squares is two-storied open space that was processed with glass curtain wall through P.F.W construction method and water space and green landscape flows naturally through the inside and outside of the building. The elevation for business area was planed by using a vertical louver so that radiant heat control is possible for each season, and for the general building plan including natural ventilation, lighting, the top floor garden, the concept of Eco-Energy was introduced aiming to offer an business area with high efficiency. About the internal path of flow, a space to look down was planed and also a sky park in connection with a park and a broad outlook created an open atmosphere. Gwangju Jeonnam branch office of Korea National Housing Corporation, designed based on this symbolism, will give urbane infra structures and events and play a role as a connector with the local society.

1. T24 COLOR LOW-E PAIRED GLASS(ONE-SIDE HALF TEMPERED)
2. T24 COLOR LOW-E PAIRED GLASS(DOUBLE-SIDES HALF TEMPERED)
3. T3 ALUMINUM SHEET
4. T12 CLEAR TEMPERED GLASS (P.F.W SYSTEM)
5. T24 COLOR LOW-E PAIRED GLASS (ONE-SIDE AND DOUBLE-SIDES HALF TEMPERED)
6. T12 TEMPERED GLASS

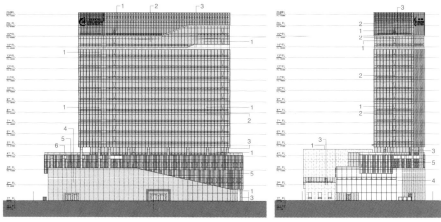

FRONT ELEVATION

LEFT ELEVATION

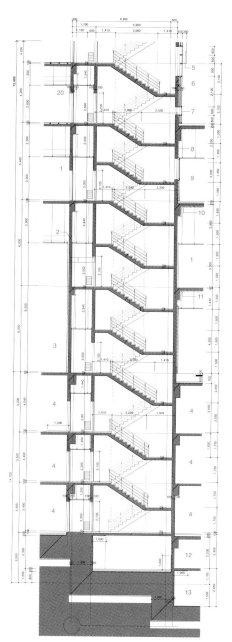

STAIR CROSS SECTION (LOW-RISE)

1 CORRIDOR
2 S.T RETAINING WALL OPEN(300X600)
3 PUBLIC INFORMATION EXHIBITION HALL
4 UNDERGROUND PARKING LOT
5 ALUMINUM CAP
6 ROOF GARDEN
7 T24 COLOR LOW-E PAIRED GLASS(ONE-SIDE HALF TEMPERED)
8 T60 GLASSWOOL
9 OUTDOOR DECK
10 T105 HEAT INSULATION
11 LIGHT GAUGE STEEL CEILING RIB / ALUMINUM CEILING PANEL
12 HEAVY - WATER EQUIPMENTS MANAGEMENT(PIT)
13 HEAVY - WATER TREATMENT FACILITY
14 STAIRWAY ROOM
15 ELECTRONIC ROOM
16 LEVELING CON'C INFILLING
17 FAN ROOM
18 FIRE EXTINGUISHING GAS ROOM
19 T50 LEVELING CON'C / T0.03 P.E FILM 2FOLDS / CODDLE STONE COMPACTION (CIVIL ENGINEERING WORK)
20 GLASS SETTING BLOCK@QUARTER LOCATION

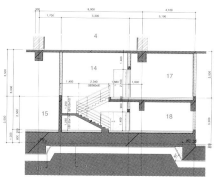

B3 STAIR ROOM SECTION

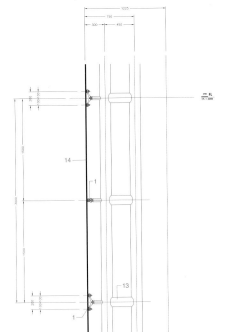

Night view of the lower exterior

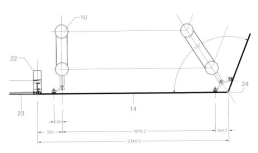

STL. STRUCTURE MODULE

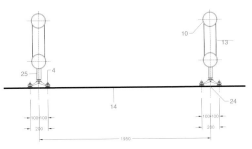

SECTION B

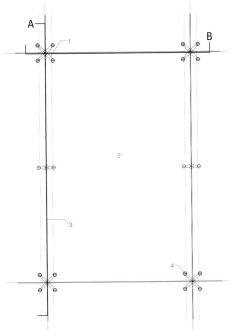

SECTION A

GLASS CURTAIN WALL PLAN

Night view of the upper exterior

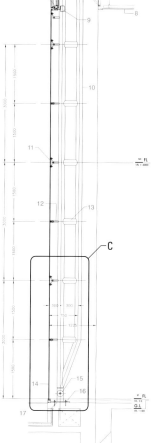

1. S'STL. SPIDER BRACKET
2. T12, TEMPERED GLASS ON ELEVATION
3. FIELD SEAL W/STRUCTURAL SILICONE SEALANT ON ELEVATION
4. POINT FIXING WALL SYSTEM BOLT
5. T3, ALUM. PANEL, ALCAN
6. T1.5, STAINLESS STL. PLATE
7. ANCHOR CLIP, ANGLE, 170X90XT8, 150LG @650 O.C
8. CEILING FINISH
9. STL. TUBE, 150X100XT9, MEMENT CONNECTION
10. STL. TRUSS FOR POINT FIXING WALL 165.2DIAXT10, PVDF FINISH
11. 165.2DIAXT10, STL. ROUND PIPE
12. 50DIA. STL COUUPLER
13. 139.8DIAXT4.5 STL. ROUND PIPE
14. T12, TEMPERED GLASS
15. 30DIA. GALV. STL. PIN
16. STL. BASE PLATE, 490X490XT15
17. 2-STL, PLATE, T12
18. FLOOR FINISH
19. GROUTING MORTAR
20. BREAK FORMED STS FRAMING, CONT.
21. M20 CHEMICAL ANCHOR 4NOS PER EACH CLIP
22. EXTRUDED ALUM. REINFORCEMENT TUBE 120X100XT3
23. T24 PAIR GLASS
24. FIELD SEAL W/STRUCTURAL SILICONE SEALANT
25. 50DIA. STL COUPLER

DETAIL C

CURTAIN WALL PARTIAL SECTION

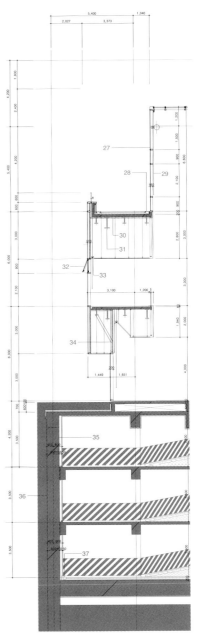

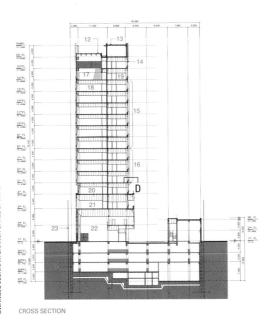

CROSS SECTION

OUTER WALL CROSS SECTION(LOW-RISE)

1   T27.2XT3 SST. PIPE
2   SST. 2L - 40X40XT5
3   T24 COLOR LOW-E PAIRED GLASS
4   T12.3 TEMPERED LAMINATED GLASS
5   T48.6XT2.5 SST. PIPE
6   T12.3 TEMPERED LAMINATED GLASS
7   WATERPROOFING CAULKING
8   INS. PL - 1/W 4-M20 CHEMICAL
9   T48.6XT2.5 SST. PIPE
10 SST. 2L - 40X40XT5
11 SST. PL - T12
12 T24./6 COLOR LOW-E PAIRED GLASS
13 T126 PLAIN CON´C, MECHANICAL PLASTERING (#8-150X150 WIREMESH) / T21 PROTECTIVE MORTAR / T3 MEMBRANE WATERPROOFING / STRUCTURAL SLAB, STEEL TROWEL FIN.
14 ALUMINUM GRILL
15 RENTAL OFFICE ZONE
16 OFFICE ZONE
17 INTERIOR GARDEN
18 RENTAL OFFICE
19 TOILET
20 MIDDLE CONFERENCE ROOM
21 GRAND CONFERENCE ROOM
22 HALL
23 SITE BOUNDARY LINE
24 SLURRY WALL (CIVIL ENGINEERING WORK)
25 T50 LEVELING CON´C / T0.03 P.E FILM 2FOLDS / CODDLE STONE COMPACTION (CIVIL ENGINEERING WORK)
26 STEEL FIREPROOFING SHUTTER INSTALL (ELECTRONICALLY POWERED)
27 T24 COLOR LOW-E PAIRED GLASS (ONE-SIDE HALF TEMPERING)
28 GRANITE TRENCH COVER INSTALL (W=300)
29 ROLL SCREEN (EXTRA WORK)
30 T105 HEAT INSULATION
31 FIREPROOF COVERING (TYP)
32 ELECTRONICALLY-POWERED WINDOW
33 STEEL CURTAIN BOX
34 C - 150X50X20X3.2@900
35 WATER-BASED PAINT / T50 ASBESTOS-FREE LIGHT-GAUGE COMPOSITE PANEL / T150 SPACE
36 CIVIL ENGINEERING RETAINING WALL
37 OPEN TRENCH INSTALL(W=200)
38 T25 FIRE STOP FOAM, T1.6 STEEL PLATE (FOR BETWEEN-FLOORS FIREPROOFING SECTION)
39 T76 PLAN CON´C, COTTON-GINNING TREATMENT(#8 - 150X150 WIREMESH) / T21 PROTECTIVE MORTAR / T3 MEMBRANE WATERPROOFING / STRUCTURAL SLAB, STEEL TROWEL FIN

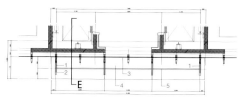

4F CANOPY SECTION D

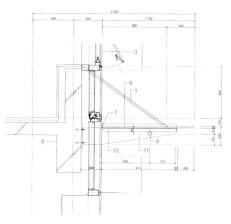

CANOPY SECTION E

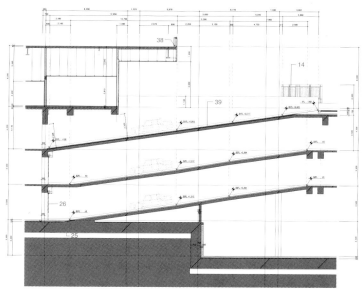

RAMP CROSS SECTION

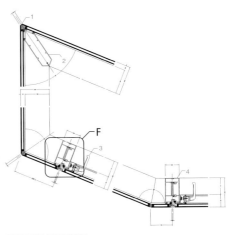

CORNER MULLION SECTION

1 FIELD SEAL W/STRUCTURAL SILICONE SEALANT
2 EXTRUDED CORNER MULLION
3 THREE POINT MULTI LOCK HANDLE
4 STRUCTURE MULLION
5 GLASS SETTING BLOCK@QUARTER LOCATION
6 T24 PAIR GLASS
7 165.2DIAXT10 STL. ROUND PIPE
8 FIELD SEAL W/SILICONE SEALANT
9 50DIA. STL. COUPLER
10 139.8DIAXT4.5 STL. ROUND PIPE
11 POINT FIXING WALL SYSTEM BOLT
12 PRIMARY STL. ANCHOR CLIP, CHANNEL 200X20XT8, 300LG
13 FIELD WELD AFTER FINAL ALIGNMENT

GLASS HORIZONTAL SECTION

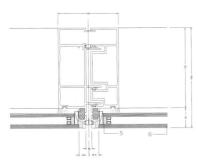

DETAIL F

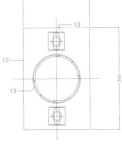

SECTION G

The lobby on the 1st floor

1 ROOF GARDEN
2 TOP WINDOW
3 RESTING ROOM
4 SMALL CONFERENCE ROOM
5 GENERAL AFFAIRS SECTION
6 LABOR UNION OFFICE
7 MIDDLE CONFERENCE ROOM
8 HVAC ROOM
9 GRAND CONFERENCE ROOM
10 OUTDOOR DECK
11 GENERAL MANAGER'S ROOM
12 STAGE
13 HALL
14 SERVER ROOM
15 BROADCASTING ROOM
16 AGENTS ROOM
17 FEMAIL EMPLOYEE'S ROOM
18 6M PASSENGER'S ROAD
19 20M ROAD
20 CAR MAIN ENTRY
21 LARGE SIZE CAR PARKING
22 UNDERGROUND PARKING LOT ENTRY
23 TEACHER'S ROOM
24 SLEEPING ROOM
25 CHILDREN-CARE ROOM
26 NURSING ROOM
27 MDF ROOM
28 CONTROL ROOM
29 RESTING HALL
30 MAIN ENTRANCE
31 PUBLIC INFORMATION EXHIBITION HALL
32 NIGHT-DUTY ROOM
33 ALMIGHTY TESTER ROOM
34 TESTING BUILDING
35 SUB-ENTRANCE
36 PUBLIC INFORMATION OFFICE (BANK)
37 SAFE

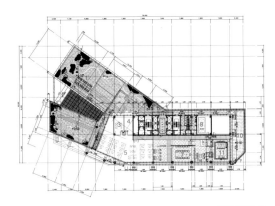

4TH FLOOR PLAN

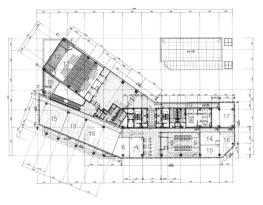

3RD FLOOR PLAN

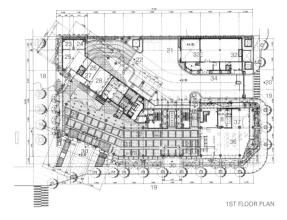

1ST FLOOR PLAN

>> COMPETITION_GWANGJU JEONNAM BRANCH OFFICE OF KNHC

광주전남지역본부 신축사옥 현상설계경기

COMPETITION_GWANG JU JEONNAM BRANCH OFFICE OF KNHC

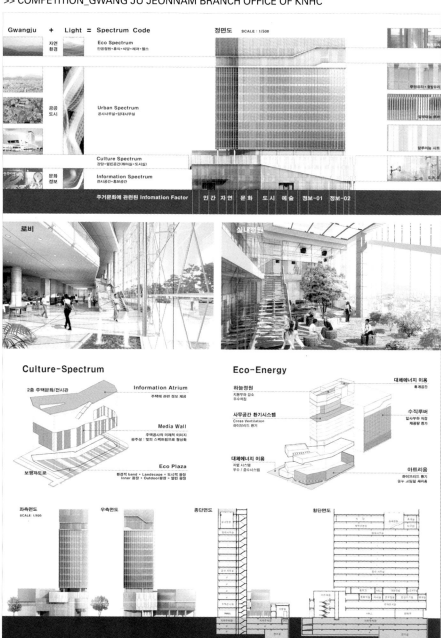

\>\> OFFICE

DAEGU GYEONGBUK BRANCH OFFICE OF KNHC

Location Dowon-dong, Dalseo-gu, Daegu, Korea **Site Area** 6,499m² **Building Area** 2,430m² **Total Floor Area** 24,859m² **Building Scope** B2–13F **Structure** SRC(Suspension) **Parking** 264 Cars **Exterior Finish** T24 Color Low-E Pair Glass, T17.52 Tempered Laminated Glass, High Strength Aluminum Sheet, High Density Wood Panel **Completion** 2007 **Architecture Design** Lee Jong-chan, Sung Jin-yong, Park Ki-seong, Lee Seung-youn | WONYANG Architectural Design Group **Design Team** Park Geun-u, Jang Sun-taek, Jeong Ju-chan, Kim Dae-hong, Park Jeong-man, Kim Jong-su **Construction** DAELIM Industrial Co., Ltd. **Photographer** Cho Tae-ryong

Daegu Gyeongbuk branch office of Korea National Housing Corporation can be summarized with the title "Clearness". This meaning has more positive and optical meaning than a similar word "openness". The architect considered a view from the lower residential areas, and defined the spacious connection from the green/theme park which encourages people to interact with others to the convenience facilities on the lower parts as "clearness". The over raised upper parts creates a vertical and transitional space, keeping distance from the lower parts. With the construction of Sky Garden, the building could have an environment-friendly space. The structure of the building has aesthetic and tectonic factors which bring out the beauty of the building. The steel structure of the top part serves to support the weight of the upper part while being an aesthetic factor creating an dynamic tension in the building. The upper part was separated from the residential area and the gap between the building and the residential area was filled up with the culture park, considering the neighborhood and privacy of the residents. Through the attempts, the architect tried to create an interactional architecture to be used as a space of communication with the neighborhood. The open lounge garden which connects the two office floors provides a feeling of refreshment, the clean air, and better working environment to the employees who work in the office. The exterior spaces including a public square, a small garden, flower beds that are open for visitors, and a ground fountain were decorated as a vibrant and friendly park for not only the employees but also people in the community. Through this "clearness", they intends to clear all the barriers which block between human(citizen) and human(employers), nature and human, and the city and the buildings.

Night view of the front elevation

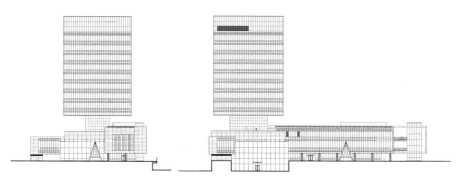

LEFT ELEVATION

REAR ELEVATION

View from the rear side

Top light located in the rear side

1 T30 GRANITE BURNER FLARING
2 HIGH STRENGTH AL. SHEET
3 T12 COLOR TEMPERED GLASS / W/ SHATTERPROOF FILM
4 Ø 89.1X3.2 / FLUOROCARBON RESIN COATING
5 Ø165.2X4.5 / FLUOROCARBON RESIN COATING
6 LIGHT GAUGE STEEL CEILING RIB / T9 ROCK WOOL SOUND-ABSORBING CEILING PLATE
7 ALL-KINDS TESTER
8 T3 SPECIAL REINFORCED FLOORING / CON'C SLAB SURFACE TREATMENT
9 T30 GRANITE BURNER FLARING (MACHEON STONE)
10 T30 GRANITE RUBBING FIN.

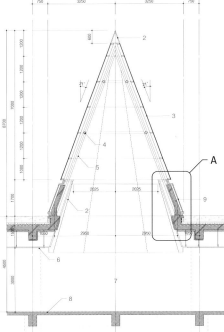

TOP LIGHT SECTION

DETAIL A

Lower part of the right elevation

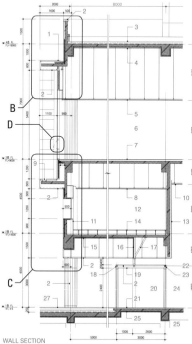

WALL SECTION

1. HIGH STRENGTH AL. SHEET / T40 GLASS WOOL
2. HIGH STRENGTH AL. SHEET
3. TOP FIN (LANDSCAPING WORK) / MIN. T70 PLAIN CON'C (SLOPE) / #8-150X150 WIREMESH / T27 CEMENT MORTAR / T3 RUBBER ASPHALT MEMBRANE WATER-PROOFING / STEM CON'C STEEL TROWEL FIN.
4. T105 HEAT INSULATION
5. LIGHT GAUGE STEEL CEILING RIB / T9.5 GYPSUM BOARD 2PLY / VINYL PAINT
6. STAFF EXERCISE FACILITY
7. WOOD FLOORING(FOR GYM) / CEMENT MORTAR
8. LIGHT GAUGE STEEL CEILING RIB / T9 ROCK WOOL SOUND-ABSORBING CEILING PLATE
9. GUTTER
10. LIGHT GAUGE STEEL CEILING RIB / GALVANIZED STEEL PLATE / WOODEN-PATTERN SHEET
11. T12.5 GYPSUM BOARD 2PLY / WATER-BASED PAINT
12. CENTRAL CONTROL ROOM
13. T10 HIGH DENSITY WOOD PANEL
14. T3 CONDUCTIVE VINYL TILE / T300 ACCESS FLOOR
15. T75 HEAT INSULATION
16. □-100X100X7 STL. PIPE
17. L-75X75X6
18. □-100X100X7 STL. PIPE / W/THK1.5X250 STAINLESS STEEL
19. COLOR ALUMINUM CEILING CEILING PLATE
20. WINDBREAK ROOM
21. T20 MARBLE RUBBING FIN. / CEMENT MORTAR
22. □-100X100X7 STL
23. T1.5X150 STAINLESS STEEL
24. LOBBY
25. T55 HEAT INSULATION
26. BASEMENT PARKING LOT
27. T30 GRANITE RUBBING FIN. / T70 CEMENT MORTAR / STEEL TROWEL FIN. / MIN. T70 PLAIN CON'C (SLOPE) / #8-150X150 WIREMESH / T27 CEMENT MORTAR / T3 RUBBER ASPHALT WATERPROOFING / STEM CON'C STEEL TROWEL FIN.
28. T24 PAIR GLASS (6TCOL+12A+6TCL)
29. STRUCTURAL W/NORTON TAPE
30. ALUM. EXTRUDED CORNER PIECES
31. SANTOPLANE WEATHER STRIP GASKET
32. L-40X40X3@900
33. T2 ALUMINUM SHEET
34. SILICONE SEALANT
35. 10X10 CAULKING
36. T20 MORTAR / WATER-BASED PAINT
37. Ø13 IRON BAR (L=600)
38. STEEL TROWEL FIN. / MIN. T70 PLAIN CON'C (SLOPE) / #8-150X150 WIREMESH / T27 CEMENT MORTAR / T3 RUBBER ASPHALT MEMBRANE WATERPROOFING / STEM CON'C STEEL TROWEL FIN.
39. T10 HIGH DENSITY WOOD PANEL / □-50X50X2.3
40. T60 HEAT INSULATION
41. HIGH STRENGTH AL. SHEET / □-50X50X2.3
42. SST GUTTER
43. 15X15 SILICONE CAULING W/ BACKER ROD
44. RING FOR CLEANING(@3600, 2 PTS PER A SPAN)
45. T6 SST PL(W:60)
46. Ø19 SST RING(INTERNAL DIAMETER : 100)
47. T24 CLEAR LOW-E PAIRED GLASS
48. T0.7 GALV. CORRUGATED STEEL PLATE / FLUOROCARBON RESIN COATING
49. T1.6 SST. PL

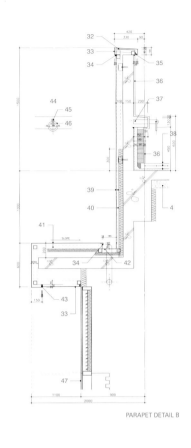

PARAPET DETAIL B

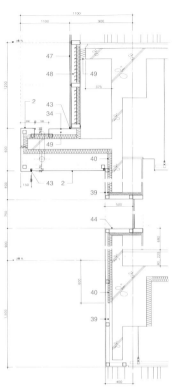

WALL DETAIL C

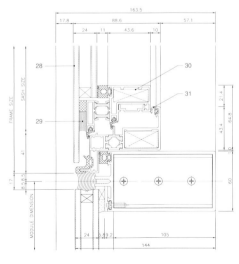

MULLION DETAIL D

Lower part of the front elevation

1. FIELD SEAL W/STRUCTURAL SILICONE SWALANT
2. T24 PAIR GLASS
3. STL.TRUSS FOR POINT FIXING WALL 165.2DIAXT10, ON ELEVATION
4. 138.9DIAXT4.5, ROUND PIPE
5. T12, TEMPERED
6. T10, STAINLESS STL. FLAT BAR
7. GROUND MORTAR
8. FLOOR FINISH
9. STL. BASE PLAATE, 490XT15
10. T2 ALUM. SHEET
11. T12 PAIR GLASS(BY OTHER'S)
12. ROLL BLIND(BY OTHER'S)
13. FIELS SEAL W/SILICONE SEALANT AND BACKER ROD, CONT
14. T20 STL. RIB
15. M20 CHEMICAL ANCHOR 4NOS PER EACH CLIP(BY OTHER'S)
16. WOOD PANEL
17. S'STL. SPIDER BRACKET
18. STL. PLATE, 95X50XT6
19. T12 TEMPERED GLASS
20. STL. PLATE, 267X197XT12 BOTH SIDE
21. STL. PLATE, 150X140XT20
22. M20 CHEMICAL ANCHOR 4NOS PER EACH CLIP
23. T12 FLAT BAR
24. STL. BASE PLATE 530X530XT15
25. GALV. STL. BOLT 30 DIA

OFFICE
DAEGU GYEONGBUK BRANCH OFFICE OF KNHC

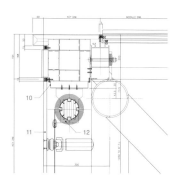

PURLIN DETAIL E

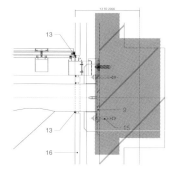

FINISH DETAIL F

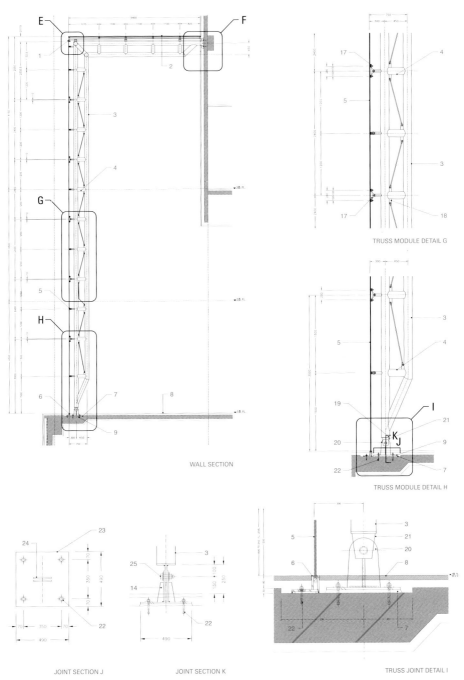

Staircase viewed from the rear side

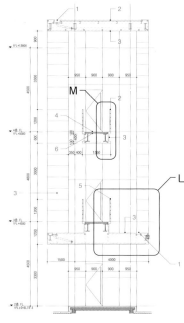
WALL SECTION

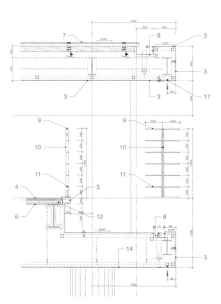
CORRIDOR PARTIAL DETAIL L

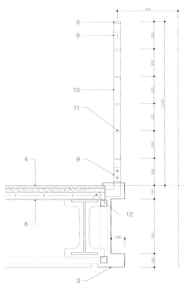
CORRIDOR PARTIAL DETAIL M

1. GUTTER
2. ALUMINUM EXTRUDED FRAME
3. HIGH STRENGTH AL. SHEET
4. T30 GRANITE RUBBING FIN. / CEMENT MORTAR / LIQUID WATERPROOFING TYPE 1
5. T12X50 SST PL
6. T50 PLAIN CON'C / #8-150X150 WIREMESH / THK6 STL. PL.
7. ALUMINUM EXTRUDED FRAME / T100 HEAT INSULATION / PURLIN
8. SST GUTTER
9. SST. FB-50X12T
10. SST. FB-50X12TX2EA
11. SST. FB-38X9T
12. RIB PL. -6@1,000
13. HIGH STRENGTH AL. SHEET / □- 50X50X2.3
14. ALUMINUM CEILING MATERIAL
15. T24 CLEAR / LOW-E PAIRED GLASS
16. T2 AL. EXTRUDED BAR
17. 15X15 SILICONE CAULKING / W/ BACKER ROD
18. HIGH STRENGTH AL. SHEET / FRP COATING WATERPROOFING
19. SILICONE SEALANT
20. BABY PLAY ROOM
21. T1.2 GALV. STL. / URETHANE PAINT
22. CAULKING (10X10)
23. T24 CLEAR LOW-E PAIRED GLASS
24. PORCELAINOUS TILE / CEMENT MORTAR / T10 PLAIN CON'C / MEMBRANE WATERPROOFING / MORTAR
25. T9.5 GYPSUM BOARD 2PLY / VINYL PAINT / LIGHT GAUGE STEEL CEILING RIB

Left part of the wall viewed from the front side

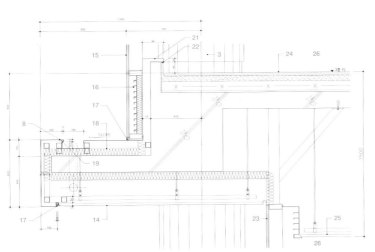

FLOOR PARTIAL SECTION

The lobby on the 1st floor

1 LABORATORY	12 ATRIUM
2 ALL-KINDS TESTER	13 CENTRAL CONTROL ROOM
3 OFFICE	14 BIDDING ROOM
4 PARKING LOT	15 KITCHEN
5 NIGHT DUTY ROOM	16 DINING HALL
6 STORAGE	17 SQUASH TENNIS
7 BANK BRANCH OFFICE	18 TABLE TENNIS COURT
8 LOBBY	19 STAFF EXERCISE FACILITY
9 BABY PLAY ROOM	20 LOCKER ROOM(MEN)
10 COMMUNITY SPACE	21 GRAND CONFERENCE
11 RESTING LOUNGE	ROOM

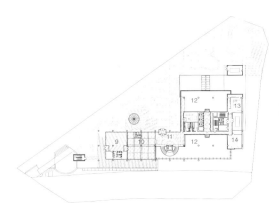

2ND FLOOR PLAN

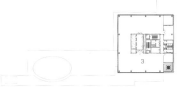

8TH FLOOR PLAN

4TH FLOOR PLAN

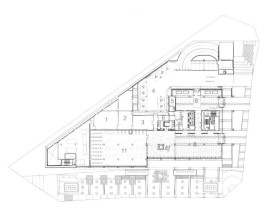

1ST FLOOR PLAN

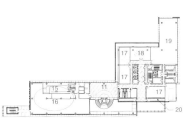

3RD FLOOR PLAN

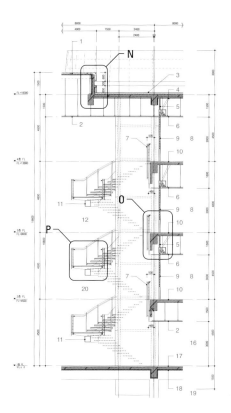

LOBBY & STAIR SECTION

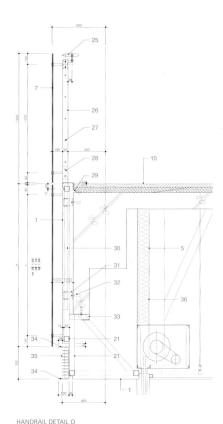

HANDRAIL DETAIL O

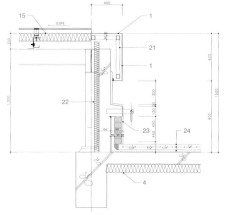

ROOF DETAIL N

1 HIGH STRENGTH AL. SHEET
2 T9.5 GYPSUM BOARD 2PLY / VINYL PAINT
3 TOP FIN. LANDSCAPING WORK / MIN. T70 PLAIN CON'C (SLOPE) / #8-150X150 WIREMESH / T27 CEMENT MORTAR / T3 RUBBER ASPHALT MEMBRANE WATERPROOFING / STEM CON'C STEEL TROWEL FIN.
4 T105 HEAT INSULATION
5 T100 GLASS WOOL
6 LIGHT GAUGE STEEL CEILING RIB (M-BAR) / T9.5 GYPSUM BOARD 2PLY / VINYL PAINT
7 T10 TEMPERED GLASS
8 RESTING LOUNGE
9 FIREPROOF SHUTTER
10 T30 GRANITE RUBBING FIN. / T50 CEMENT MORTAR
11 STL. / FLUOROCARBON RESIN COATING
12 OPEN (ATRIUM)
13 T2 STAINLESS LOUVER
14 T12 TEMPERED GLASS
15 ALUMINUM EXTRUDED FRAME / T100 HEAT INSULATION / PURLIN
16 EXHIBITION LOBBY
17 T20 MARBLE RUBBING FIN. / T50 CEMENT MORTAR
18 T55 HEAT INSULATION
19 BASEMENT PARKING LOT
20 EXHIBITION LOBBY
21 □ - 50X50X2.3
22 T60 HEAT INSULATION
23 T20 MORTAR / WATER-BASED PAINT
24 STEEL TROWEL FIN. / MIN. T70 PLAIN CON'C (SLOPE) / #8-150X150 WIREMESH / T0.05 PE FILM 2-FOLDED SPREADING / T27 CEMENT MORTAR / T3 RUBBER ASPHALT MEMBRANE WATERPROOFING / STEM CON'C STEEL TROWEL FIN.
25 AL. EXTRUDED BAR(FLUOROCARBON COATING)
26 SST. FB - 50X12TX2EA
27 Ø10 SST. ROD BAR@200
28 2-M9 SST. BOLT
29 SST. FB-2.3X30@450
30 SST. FB-50X12T
31 L-60X90X5
32 1-Ø9 SET ANCHOR
33 L-60X60X5
34 T10X100 SST.PL
35 T3X60 SST. PL@28
36 T1.2 GALV. STL.
37 SST. FB-38X9T
38 1-M9 SST. BOLT
39 PL-12T (100X150)
40 D10 (2EA)
41 Ø50.8 SST PIPE
42 Ø9 SST CIRCULAR BAR
43 PL-6
44 T40 GRANITE RUBBING FIN. (C-BLACK) / T40 CEMENT MORTAR / W/ METALLATH
45 PL-12
46 PL-6 STL. / FLUOROCARBON RESIN COATING
47 PL-12 STL. / FLUOROCARBON RESIN COATING
48 □ - 450X450X12 STL. / FLUOROCARBON RESIN COATING
49 AL. SHEET

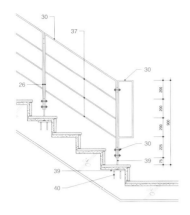
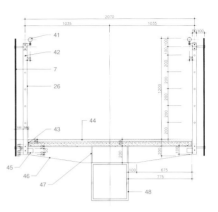

STAIR & HANDRAIL DETAIL P

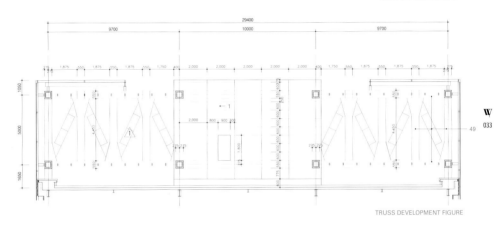

TRUSS DEVELOPMENT FIGURE

Steel truss of the top part

>> COMPETITION_DAEGU GYEONGBUK BRANCH OFFICE OF KNHC

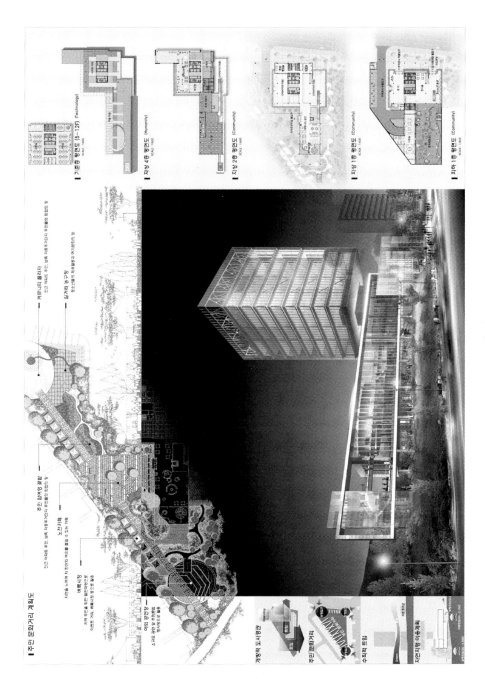

>> OFFICE

CULTURE & CONTENT CENTER

Location Sangam-dong, Mapo-gu, Seoul, Korea Site Attribute District Unit Plan, General Commerce Site Area 6,611m² Building Area 3,963m² Total Floor Area 43,610m² Building Coverage Ratio 59.95% Floor Area Ratio 393.83% Building Scope B4–14F Structure SRC, RC Exterior Finish T24 Color Pair Glass, T12 Clear Tempered Glass, Wood Panel, Aluminium Composite Panel Design Period 2003.12–2007.3 Construction Period 2004.12–2007.3 Architecture Design Lee Jong-chan, Sung Jin-yong, Park Ki-seong, Lee Seung-youn | WONYANG Architectural Design Group Photograph WONYANG Architectural Design Group

WALKING THE D.M.S(DIGITAL MEDIA STREET) It was a subject to set an outstanding landmark since C3 in the D.M.C(Digital Media City) site symbolizes Digital Media and locates at the entrance of the D.M.C complex. City, architecture, and interactions to utilize the 20m long pedestrian road were the most crucial key words in this project. Two separate spaces (Cultural Contents Center, Video Archive Center) within this center brought us some interesting subjects such as spacial connection, partition by use, and space allocation.

CITY MEETS ARCHITECTURE AND ARCHITECTURE MEETS HUMAN BEINGS Because the C3 site has respectively 20m and 10m long pedestrian roads, the site had to be easily accessible from the pedestrian roads and widely open. Thus, the architects attempted to harmonize the exterior with the internal spaces through the 16m high spacious lobby. The pedestrian roads and two Sunkens connected to the roads lead people enter the underground movie theater and provide them with Ssamji lounge area as well as a shelter just for an emergency case. They allocated the cultural contents support facilities to the 3rd and 4th floors to minimize effects to the surroundings and constructed a lounge terrace in each facility. In addition, they applied the 8-9m optimal span to create pleasant working conditions. The sky court on the 5th floor as a park in the city center, spacious public space, and the huge lounge atrium constructed from 9th floor up to 14th floor bring us a cubic communication space and the dynamic cultural contents center.

NATURE MINGLES WITH ARCHITECTURE AND NATURE MINGLES WITH HUMAN BEINGS The cafeteria and the multi experiences center designed with D.P.G(Dot Point Glazing) technology minimizes the boundary between the internal space and exterior space. Since it faces the Media Street, it promotes people to utilize the pedestrian roads and their interactions. "Bium" spreaded on four floors creates an unique identity with the spacious space, independent location and the horizontal features. The red cedar panel which was used to finish the exterior walls of the Multi Video Archive Center emphasizes the symbolic meaning of storage building while creating a comfortable atmosphere with the AL panel which was interpreted in a modern way as a structural foundation. The clear atrium and the Sky court on the roof and the Bium spreaded on four floors create an environment-friendly atmosphere by letting the natural sunlight and air into the office as well as the central core finished with four mega columns and the extruded cement panels. It will play a central role as hub of D.M.C due to the cheerful working conditions.

General view of the facade

1. T20 WOOD PANEL ("A" TYPE)
2. T20 WOOD PANEL ("B" TYPE)
3. T20 WOOD PANEL ("C" TYPE)
4. T30X150(@150) WOOD LOUVER ("B" TYPE)
5. 150X20 RED CEDAR / OIL STAIN (APP. COLOR) / BUILDING PAPER / T12 WATERPROOF PLYWOOD / 50X50X2.3 STL PIPE
6. T2 ALUMINUM SHEET
7. T75 HEAT INSULATION
8. 150X20 RED CEDAR / OIL STAIN (APP. COLOR)
9. BUILDING PAPER
10. T12 WATERPROOF PLYWOOD
11. ㄷ-57X75X75X6T ST'L PLATE
12. ㅁ-50X50X2.3 STL PIPE
13. RED CEDAR / OIL STAIN (APP. COLOR)
14. LOBBY CEILING FINISHING LINE
15. GALVANIZED SCREWPIECE (FOR WOODWORK)
16. T12 TEMPERED GLASS
17. T7 SST F.B
18. L-110X75X7 PROCESS
19. T7 SST F.B PROCESS

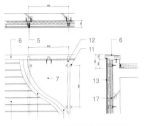

WOOD WALL SECTION A

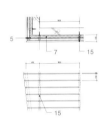

WOOD WALL SECTION B

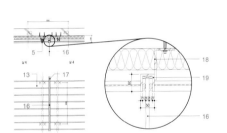

WOOD WALL SECTION C

WOOD WALL SECTION D

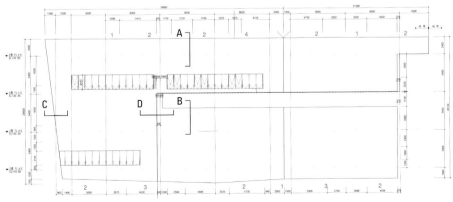

WOOD WALL DEVELOPMENT FIGURE

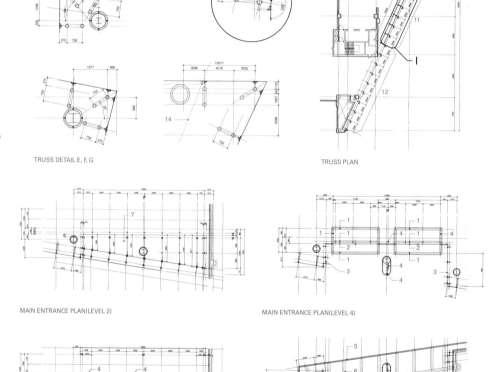

TRUSS DETAIL E, F, G

TRUSS PLAN

MAIN ENTRANCE PLAN(LEVEL 2)

MAIN ENTRANCE PLAN(LEVEL 4)

MAIN ENTRANCE PLAN(LEVEL 1)

MAIN ENTRANCE PLAN(LEVEL 3)

The truss structure of the left elevation

1 H-200X200X8X12(BY OTHERS)
2 FRAME & GLASS(BY OTHERS)
3 T12 TEMP. SPG GLASS(INNER SAFETY FILM)
4 Ø165.2X10T STL. PIPE
5 CANOPY END LINE (BY OTHERS)
6 H-BEAM(BY-OTHERS)
7 T17.52 LAMI. TEMP. GLASS
   (10CL.T/P+1.52PVB +6CL.T/P)
8 PIPE TRUSS-TYPE-1
9 PIPE TRUSS-TYPE-2
10 PIPE TRUSS-TYPE-3
11 PIPE TRUSS-TYPE-4
12 PIPE TRUSS-TYPE-5
13 T2 ALUMINUM SHEET
14 END LINE OF UPPER WALL
15 Ø165.2X10T@2000
16 T12 ST'L PLATE
17 Ø30(M30)
18 T30 NON-SHRINK MORTAR
19 SET ANCHOR (M-16)

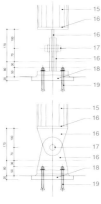

TRUSS BASE PLATE

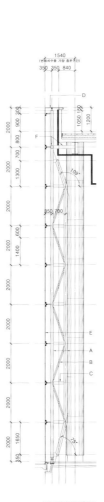

TRUSS SECTION H

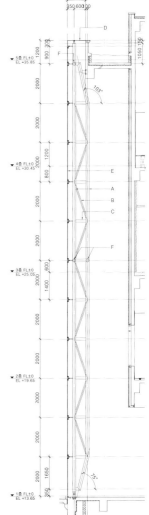

TRUSS SECTION I

The outdoor garden on the 5th floor

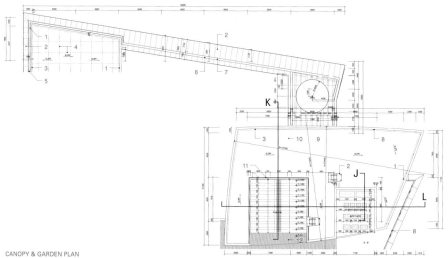
CANOPY & GARDEN PLAN

OFFICE
CULTURE & CONTENT CENTER

1. Ø100 CAST-IRON ROOF DRAIN
2. T24 ALUMINUM COMPOSITE PANEL
3. OPEN TRENCH(W=200)
4. EXPANSION JOINT(2000X2000)
5. T0.4 STS GUTTER(W=300)
6. Ø50 CAST-IRON ROOF DRAIN
7. GUTTER(W=300)
8. RING INSTALL FOR CLEANING(Ø4000)
9. EN TRENCH
10. WATERPROOFING URETHANE EXPOSED WATERPROOFING
11. HANDRAIL("C" TYPE)
12. T20 WOOD DECK
13. T2 ALUMINUM SHEET
14. T20 RED CEDAR
15. FALSE JOINT(4000X3000) / T60(MIN.) PLAIN CON'C(SLOPE (#8X150X150 WIRE MESH) / T110 HEAT INSULATION / T10 COMPRESSED EXPANDED POLYSTYRENE / T0.02 PE FILM 2-PLY / T3 MEMBRANE WATERPROOFING / STEM CON'C STEEL TROWEL FIN.
16. INSULATION("C" TYPE)
17. F.C.U. COVER
18. T3 EXPOSED URETHANE WATERPROOFING / STEM CON'C STEEL TROWEL FIN.
19. FIREPROOF COATING ON RUSTPREVENTING (3-HRS)
20. CLAY BRICK / CEMENT MORTAR / T3 MEMBRANE WATERPROOFING
21. CURTAIN BOX("A" TYPE)
22. STEEL TROWEL FIN. / T120 PLAIN CON'C (SLOPE)(#8X150X150 WIRE MESH) / T10 COMPRESSED EXPANDED POLYSTYRENE / T0.02 PE FILM 2-PLY / T3 MEMBRAN WATERPROOFING / STEM CON'C STEEL TROWEL FIN.
23. FINISH : LANDSCAPING WORK / T60(MIN.) PLAIN CON'C(SLOPE)(#8X150X150 WIRE MESH) / T10 COMPRESSED EXPANDED POLYSTYRENE / T0.02 PE FILM 2-PLY / T3 MEMBRANE WATERPROOFING / STEM CON'C STEEL TROWEL FIN.
24. T12 CLEAR TEMPERED GLASS
25. OIL STAIN ON T30 RED CEDAR / 30X30 SQUARE PIPE(@600)
26. ANCHOR BOLT / FASTNER ANGLE
27. URETHANE WATERPROOFING
28. OIL STAIN ON T30 RED CEDAR(W=600)
29. OIL STAIN ON T30 RED CEDAR(W=350)

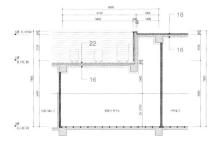

OUTDOOR GARDEN LONGITUDINAL SECTION J

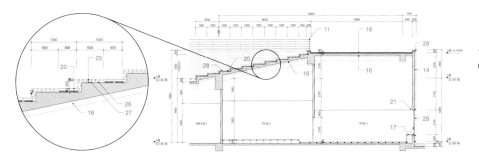

OUTDOOR GARDEN LONGITUDINAL SECTION K

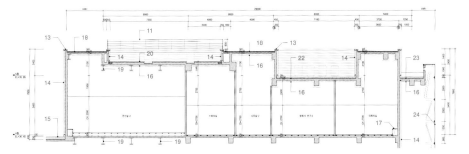

OUTDOOR GARDEN CROSS SECTION L

The canopy of the outdoor garden

1 ST'L ☐-75X45X2.3T PIPE HORIZONTAL MEMBER(RUST-PREVENTING MORE THAN TWICE) / (UP,DOWN @1000)
2 ST'L ☐-75X45X2.3T PIPE VERTICAL MEMBER(VERTICAL REINFORCEMENT)
3 T4.0 ALUMINUM COMPOSITE PANEL
4 ST'L ☐-75X45X2.3T PIPE VERTICAL MEMBER (RUST-PREVENTING MORE THAN TWICE)
5 H-BEAM(BUILDING STRUCTURE COMPONENT)
6 ST'L ☐-75X45X2.3T PIPE(RUST-PREVENTING MORE THAN TWICE) / HORIZONTAL MEMBER
7 ST'L ☐-75X45X2.3T PIPE(RUST-PREVENTING MORE THAN TWICE) / VERTICAL MEMBER
8 ST'L L-100X1000X6T 100LG BRACKET (RUST-PREVENTING MORE THAN TWICE)
9 T4.0 ALUMINUM COMPOSITE PANEL
10 CORNER PANEL INSIDE-REINFORCING PLATE
11 WATER SILICONE SEALANT W/BACK UP ROD SPONGE

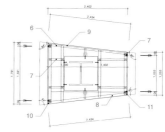

SECTION M

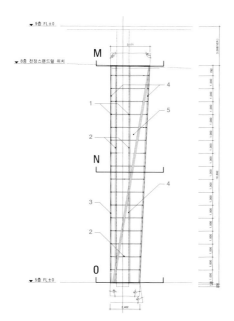

CANOPY COLUMN PLAN

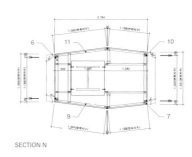

SECTION N

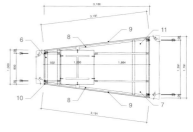

SECTION O

7TH FLOOR PLAN

1	SUNKEN
2	CONFERENCE ROOM
3	SNACK BAR
4	WAITING HALL FOR THEATER
5	TICKET BOX
6	1ST FLOOR OUTER WALL LINE
7	KOREAN FOOD
8	THEATER
9	FLOUR FOOD
10	CHINESE FOOD
11	JAPANESE FOOD
12	BEER HOUSE
13	WAREHOUSE
14	UNDERGROUND PARKING LOT
15	OFFICE
16	LOBBY
17	ROAD BOUNDARY LINE
18	10M ADJACENT ROAD
19	CAFETERIA
20	2ND FLOOR OUTER WALL LINE
21	FILM MUSEUM
22	MAIN ENTRANCE
23	BOUNDARY LINE OF ADJACENT SITE
24	CONTROL CENTER
25	CONTENTS MART
26	OPEN VACANT LAND
27	ENTRANCE FOR EQUIPMENT
28	NIGHT DUTY ROOM
29	CONTROL OFFICE
30	SUB ENTRANCE

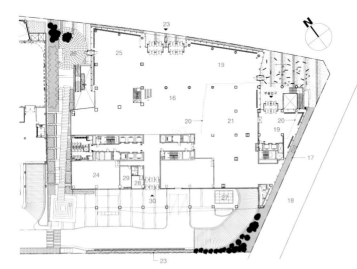

1ST FLOOR PLAN

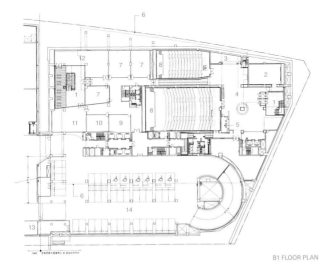

B1 FLOOR PLAN

>> COMPETITION_CULTURE & CONTENT CENTER

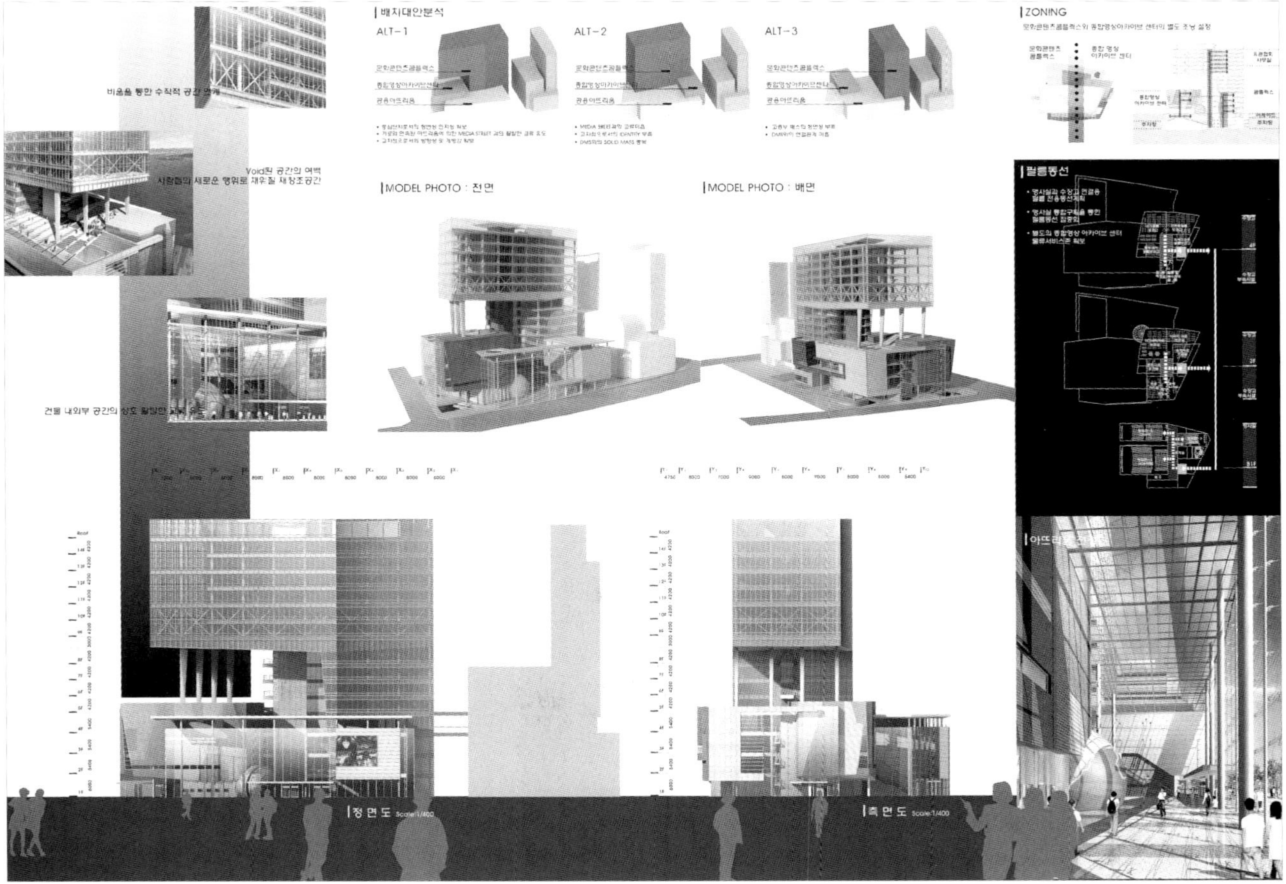
비움을 통한 수직적 공간 연계
Void된 공간의 여백
사람들의 새로운 행위로 채워질 재창조공간
건물 내외부 공간의 상호 활발한 교류 유도
배치대안분석
ALT-1
ALT-2
ALT-3
ZONING
MODEL PHOTO : 전면
MODEL PHOTO : 배면
필름동선
정면도 Scale:1/400
측면도 Scale:1/400
아뜨리움 공간

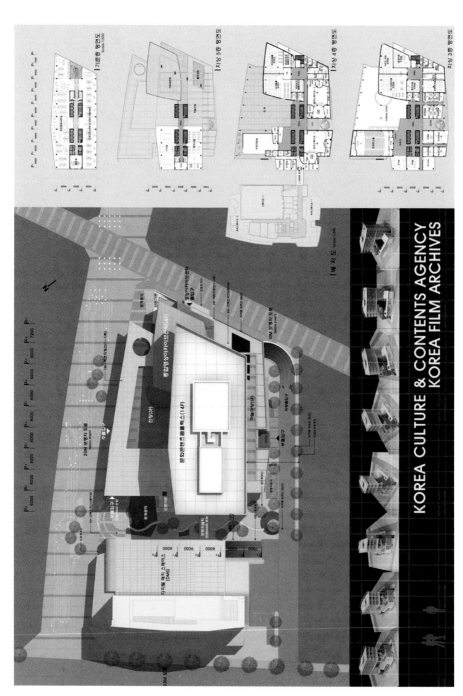

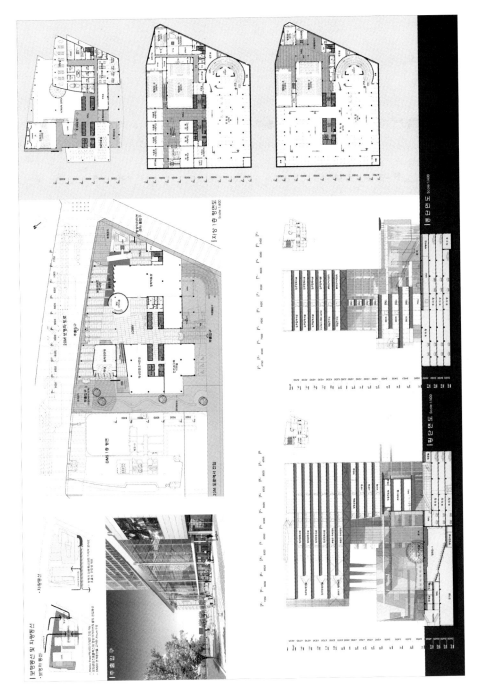

\>\> OFFICE

WOONGJIN THINKBIG

Location Munbal-ri, Gyoha-eup, Paju-si, Gyeonggi-do, Korea Site Attribute Semi Industry Site Area 6,642.8m² Building Area 3,226.93m² Total Floor Area 12,862.84m² Building Coverage Ratio 48.58% Floor Area Ratio 94.23% Building Scope B2–2F Structure SRC Exterior Finish T28 Pair Glass(Multicoating, Clear), U-Glass(Double Glazing), T20 Wood(IPE) Interior Finish T18 Wood(Mohave), Perforated Sound-Absorbing Panel, Water-Based Paint Design Period 2004.9–2005.4 Construction Period 2005.4–2007.3 Architecture Design Kim In-cheurl | Chungang University + ARCHIUM Architect Interior Design ARCHIUM Architect Design Team Jeong Yong-sik, Kang Nan-hyeong Construction WOONGJIN E&C Client Woongjin Group Corporation Photographer Lee Ki-hwan

The Woongjin Thinkbig was built to look like a rock, according to the guidelines of Paju Book City. As long as they agreed to the principles of the city, their case should not be regarded as being under the control by the authority. The building could be constructed out of the corny theory which takes 'forming' as a start of construction. You better imagine a light, feather-like rock floating over the field of reeds, instead of a heavy rock stuck in the ground. They wanted to get a transparent and light impression of the building.

A publishing company is a place which makes books. Books are made by people. Therefore, a publishing company should be a place for people who make the books. And we can surely say that Woongjin Thinkbig is definitely that kind of place. Nothing should be defined as a perfect place for a publishing company. What they need is just an open space. So there are more open spaces but less closed spaces in the building. To connect a space with another space, they removed or lowered the partitions, or they installed a transparent partition.

When we have a form and contents, it is time to think of how to fill the space. Because the meaning equals the content of the space. There is a huge court in Woongjin Thinkbig. It is a kind of outside surrounded by the inside. On one hand, without a roof, it is seen as outside the building, but on the other hand, it is seen as inside because it is actually located in the building. Outside that connects inside with an another inside should be regarded as inside since it becomes a part of the space. The spaces surrounding the court cross over and look over each other.

The space in a building is made of floors, walls and a roof. It is normally hexahedral, however, only the walls can be seen from outside the building. The floor cannot be seen because it is faced to the ground and the roof also cannot be seen because it is directed towards the sky. There is no roof in Woongjin Thinkbig. The roof was transformed into the artificial ground. We can stand or sit on the ground to look over the Han river and the field of reeds along the river, as well as the beautiful sunset because they transformed the roof into the space for break time.

They aimed to make Woongjin Thinkbig to a moderate and sophisticated building. Because a building must remain as a background to highlight people who work in the building and the books they produce in the building. An unexpected harmony of the wooden louver and Woongjin Thinkbig Collections well-arranged on the shelves brought vividness to this simple and dry space. I believe that we can refine the space better by simplifying it, not stuffing it.

General view from the west. Woongjin Thinkbig seems like a light, transparent rock embedded in the ground according to the topography.

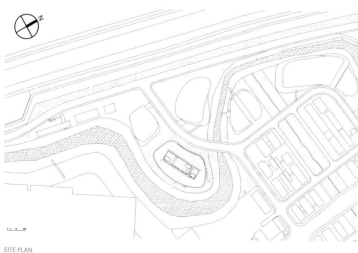
OFFICE
WOONGJIN THINKBIG
SITE PLAN

1. T25 WOOD PLANKING (IPE) / T75 PLAIN CON'C / T40 PROTECTIVE MORTAR / SURFACE COATING WATERPROOFING / INSULATION SHEET /THK150 DECK PLATE / INSULATION
2. T20 GLASSWOOL CEILING MATERIAL
3. T5 VINYL TILE FOR O.A/O.A FLOOR
4. VERTICAL LOUVER / T28 PAIR GLASS (MULTY COATING, NATURAL COLOR)
5. T5 VINYL TILE FOR O.A/O.A FLOOR
6. T16 TOPAKUSTIK
7. T28 PAIR GLASS (MULTI COATING, NATURAL COLOR)
8. T9 LAMINATED FLOORING(FOR HEAVY WALKING)
9. VERMICULITE SPRAYING
10. ASPHALT PRIMER / INFILTRATIVE WATER-PROOFING / T15 BONDING WOOD / CON'C WALL
11. OPEN TRENCH (W:100)
12. COLORCRETE / T200 PLAIN CON'C / T50 DRAIN PLATE / INFILTRATIVE WATER-PROOFING / SURFACE COATING WATER-PROOFING (COVER OVER 2,000MM FROM END LINE OF RETAINING WALL) / T60 LEVELING CON'C / PE FILM TWOFOLD / T150 CODDLE STONE COMPACTION
13. COLORCRETE / T200 PLAIN CON'C / T50 DRAIN PLATE / INFILTRATIVE WATER-PROOFING / T60 LEVELING CON'C / PE FILM TWOFOLD / T150 CODDLE STONE COMPACTION

General view from the south-west

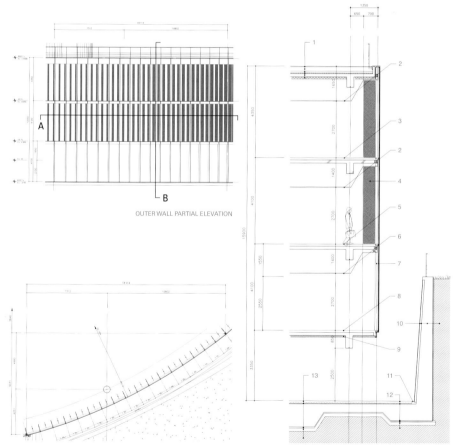

OUTER WALL PARTIAL ELEVATION

SECTION A

SECTION B

The main entrance naturally guides the visitors to the interior lobby and the courtyard.

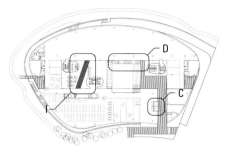

1. 30X50 ST'L PIPE @900
2. 30X40 LAUAN MDF
3. 30X40 MDF MOULDING(ONE SIDE GROOVE CHIPPING)
4. 30X30 ST'L PIPE @450
5. 30X40 LAUAN MDF @450
6. T=16MM K-TOP LINIER 28/4 / FIN: LG DECO MW201Q3
7. LIGHTING FIXTURE
8. PET 25T + K-TOP LINIER 16T
9. LINE DIFFUSER
10. THK.16 K-TOP LINIER / TYPE: 28/4M / CORE MATERIAL: T=15MM MDF / FINISH: LG DECO MW201Q3

KEY PLAN

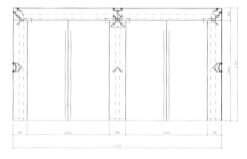

ENTRANCE ELEVATION

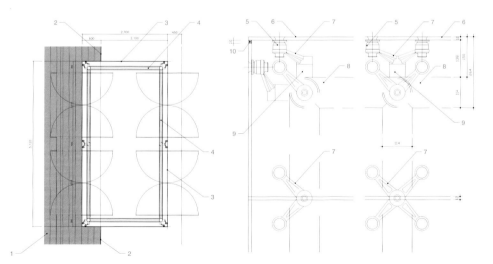

ENTRANCE PLAN C   JOINT DETAIL

The courtyard gradually revealed through the main entrance.

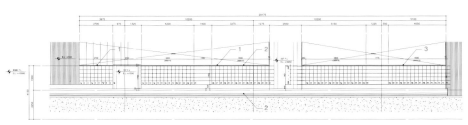

COURTYARD STAIR PLAN D

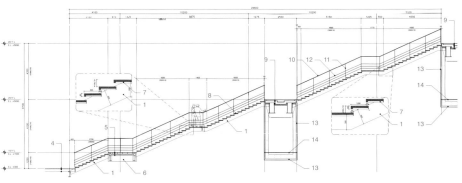

COURTYARD STAIR ELEVATION & DETAIL

1. STEEL PLATE(W300X12T @288, FLUORO-CARBON RESIN COATING
2. HANDRAIL
3. STEEL PLATE(W300X22T @288, EPOXY PAINT)
4. T30 WOOD PLANKING / T5 DAMPING RUBBER / 210X40X2.3T STEEL PLATE PROCESSING(@400, MIX PAINT, WELDING FIX)
5. T25 WOOD PLANKING / T5 DAMPING RUBBER / 40X40 WOOD FRAME / T40 PROTECTIVE MORTAR / SURFACE COATING WATERPROOFING
6. T1.2 STEEL PLATE(ELECTROSTATIC POWDER COATING)
7. T6 STEEL PLATE PROCESSING (FLUORO-CARBON RESIN COATING, WELDING FIX)
8. T25 WOOD PLANKING / T5 DAMPING RUBBER / T6 STEEL PLATE PROCESSING(EPOXY PAINT)
9. T25 WOOD PLANKING / T5 DAMPING RUBBER / 40X40 WOOD FRAME / T40 PROTECTIVE MORTAR / SURFACE COATING WATERPROOFING / CON'C SLAB / T125 SOFT WATER-BASED FOAMED INSULATION
10. HANDRAIL Ø32 SUS(VIBRATION FIN.)
11. BALUSTER SUS FB-32X12 (VIBRATION FIN.)
12. MIDDLE HANDRAIL Ø12 SUS(VIBRATION FIN.)
13. T12 TEMPERED GLASS(CLEAR, SHATTER-PROOF FILM ATTACH)
14. T31.52 LAMINATED TEMPERED GLASS (WHITE FILM, NON-SLIP)
15. T25 APP WOOD / 40X40 JOIST / T75 PLAIN CON'C / T40 PROTECTIVE MORTAR / RUBBERIZED ASPHALT MIX WATERPROOFING / THK.150 DECK PLATE / T150 SOFT WATER-BASED FOAMED INSULATION
16. COLORCRETE / T200 PLAIN CON'C / T50 DRAIN PLATE / INFILTRATIVE WATER-PROOFING / CON'C SLAB / T60 LEVELING CON'C / PE FILM TWOFOLD / T150 CODDLE STONE COMPACTION
17. T3 AL SHEET
18. DECK FINISHING LINE
19. EXPOSED CON'C(PARQUET)
20. T28 PAIR GLASS(MULTY COATING, CLEAR)
21. FILED WEATHER SILICONE SEALANT/ W / BACK UP ROD(BY OTHER'S)
22. STRUCTURAL SILICONE SEALANT / WITH NORTON TAPE
23. E.P.D.M INNER GLAZING GASKET
24. 5.5X25 LG PAN HD / SUS SCREW @350
25. 45X16 LG TRUSS HD / SUS SCREW
26. TRANSOM LINE
27. ALUM. EXTRUDED T-CLEAT
28. ALUM. EXTRUDE CLIP @350
29. THK. 28MM PAIR GLASS(8+12+8)(BY OTHER'S)

General view of the courtyard from the south-west

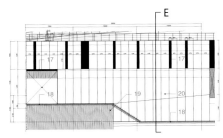

COURTYARD CURTAIN WALL ELEVATION

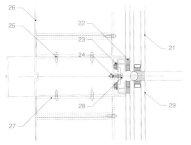

CURTAIN WALL JOINT DETAIL F

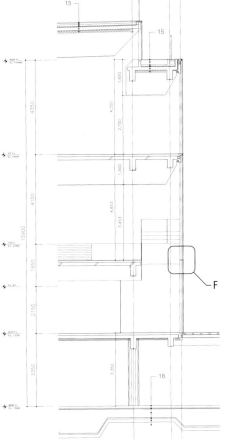

CURTAIN WALL SECTION E

General view of the courtyard from the western roof

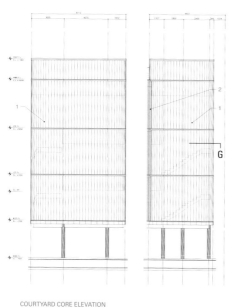

COURTYARD CORE ELEVATION

SECTION G

1   U-GLASS(DOUBLE GLAZING)
2   T28 PAIR GLASS(MULTY COATING, CLEAR)
3   U-PROFILED GLASS(VERTICAL STRIP) / (K25X60X7T GREEN COLOR & ORANGE PEEL SURFACE)
4   RUBBER PAD T20
5   WEATHER SEALANT(CLEAR COLOR)
6   AL. FRAME(HORIZONTAL MEMBER)
7   WEEP HOLES @1000/06X15LG
8   WEATHER SEALANT / BACKER-ROD ON SPONGE
9   VINYL INSERT(L=220)
10 M10X50LG / BOLT, NUT / W / WASHER
11 INTERIOR FINISH(BY OTHER'S)
12 VINYL SHEET / THK 1.0
13 ST'L ANGLE @800 / 125X180X10T.150LG
14 M12X100LG SET ANCHOR
15 THK25 IRON WOOD / WOOD JOIST / PLAIN CONCRETE / PROTECTIVE MORTAR / COMPLEX WATERPROOFIN / CON'C SLAB
16 T150 SOFT WATER-BASED FOAMED INSULATION / ACRYL PAINT ON T9.5 GYPSUM BOARD (2 PLY) / COLUMN(EPOXY PAINT)
17 THK3 EPOXY PAINT(TRANSPARENCY, CLEAR) / THK0.3 EPOXY COATING(APP. COLOR) / PLAIN CONCRETE
18 THK9.5 ACRYL PAINT ON GYPSUM BOARD(2 PLY)
19 COLUMN (EPOXY PAINT)
20 THK90 VERMICULITE SPRAYING
21 U-GLASS(SINGLE GLAZING)
22 THK60 BLACK CRUSHED STONE FILLING / THK75 PLAIN CONCRETE / THK40 PROTECTIVE MORTAR / WATERPROOFING / CON'C SLAB
23 RECESSED LIGHTING FIXTURE
24 THK1.2 OIL FLAT PAINT ON ST' PLATE
25 ST'L PLATE 300X9T EPOXY PAINT
26 H-BEAM 400X200X8X13(FIREPROOF PAINT, 1 HOUR, EPOXY PAINT)
27 COLORCRETE / THK100 PLAIN CONCRETE / THK50 DRAINAGE PLATE / INFILTRATION WATERPROOFING / CON'C SLAB
28 THK60 LEVELLING CONCRETE / PE FILM(2PLY) / THK150 CODDLE STONE COMPACTION

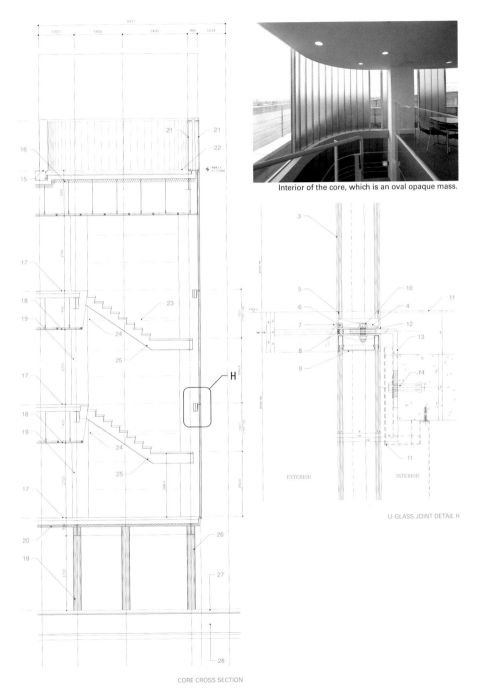

Interior of the core, which is an oval opaque mass.

CORE CROSS SECTION

U-GLASS JOINT DETAIL H

1. T25 WOOD PLANKING / T5 DAMPING RUBBER / 40X40 WOOD JOIST / T40 PROTECTIVE MORTAR / RUBBERIZED ASPHALT MIX WATERPROOFING CON'C SLAB / T125 SOFT WATER-BASED FOAMED INSULATION
2. THK3 AL. SHEET
3. SEALANT
4. TSC-GIRDER(FIREPROOF PAINT FOR 1 HR)
5. T1.6 STEEL PLATE(ELECTROSTATIC POWDER COATING)
6. STEEL PLATE(2-250X20T, FLUOROCARBON RESIN COATING ON RUST-PREVENTING)
7. ACRYL PAINT ON T9.5 GYPSUM BOARD (2PLY)
8. SYSTEM BOLT
9. T12 CLEAR TEMPERED GLASS (SHATTER-PROOF FILM ATTACH)
10. STEEL PLATE(250X22T, FLUOROCARBON RESIN COATING ON RUST-PREVENTING)
11. ST'L PIPE Ø50.8X4.0T(FLUOROCARBON RESIN COATING ON RUST-PREVENTING)
12. PLATE-12T(W=30)
13. T31.52 LAMINATED TEMPERED GLASS (WHITE FILM / NON-SLIP)
14. ST'L PIPE Ø39.8X3.6T(FLUOROCARBON RESIN COATING ON RUST-PREVENTING)
15. ST'L PLATE(350X22T, FLUOROCARBON RESIN COATING ON RUST-PREVENTING)
16. ST'L PLATE(2-350X20T, FLUOROCARBON RESIN COATING ON RUST-PREVENTING)
17. BALUSTER SUS FB-32X12(VIBRATION FIN., @1250 OR UNDER)
18. T25 WOOD PLANKING
19. HANDRAIL 32 SUS(VIBRATION FIN.)
20. ST'L PIPE Ø89.2X3.2T(FLUOROCARBON RESIN COATING ON RUST-PREVENTING)
21. BRACE (18 SEMALLOY TIE BAR)
22. T31.52 LAMINATED TEMPERED GLASS (15CL+1.52PVB+15CL, WHITE, NON-SLIP)
23. MIDDLE HANDRAIL Ø12 SUS(VIBRATION FIN.)
24. BALUSTER SUS FB-32X12(VIBRATION FIN., @1250 OR UNDER)
25. WELDING

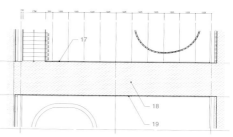

ROOF BRIDGE PLAN I

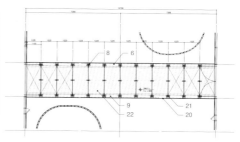

2F BRIDGE PLAN I

The exterior bridge runs above the inner corridor generated by the introduction of the glass wall.

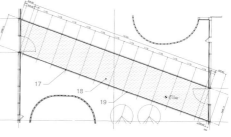

1F BRIDGE PLAN I

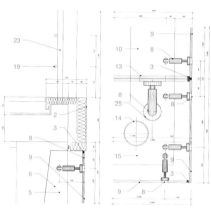

UPPER JOINT DETAIL J LOWER JOINT DETAIL K

The oval mass made of U glass is a space for the core.

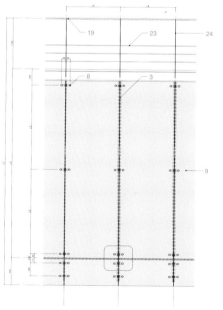

BRIDGE ELEVATION

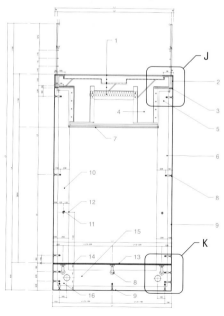

BRIDGE CROSS SECTION

1 THK28 PAIR GLASS(MULTY COATING, NATURAL COLOR)
2 T25 WOOD PLANKING
3 HANDRAIL(REFER TO EXTRA DRAWING)

General view of the roof from the west. As yet another free space connected to the courtyard, the roof presents the character of a curve and not a plane.

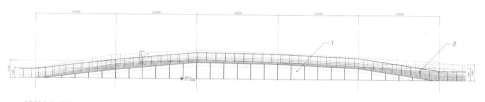

ROOF ELEVATION

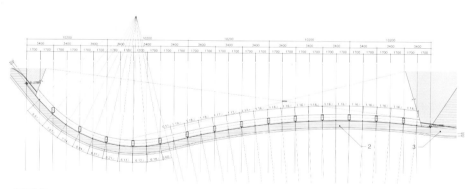

ROOF PLAN

The spaces encircling the courtyard are linked to one another by the multiple bridges.

Night view of the courtyard from the north-east

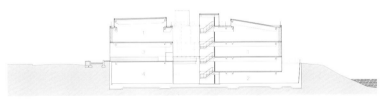

1 WORKROOM
2 PARKING LOT
3 LOBBY
4 MACHINE ROOM

LONGITUDINAL SECTION

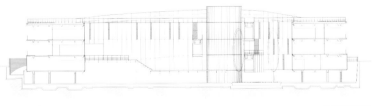

CROSS SECTION

General view of the north-western interior from the courtyard. The transparent glass allows an open view between the courtyard and the interior.

Inner corridor on the 1st floor. The interior and the exterior communicate with each other in every place.

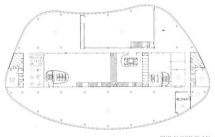

2ND FLOOR PLAN

1 PARKING LOT
2 MACHINE ROOM
3 GENERATOR ROOM
4 STORAGE
5 WORKROOM
6 CAFETERIA
7 KITCHEN
8 STUDIO
9 SEMINAR ROOM
10 DATA ROOM
11 CONFERENCE ROOM
12 HVAC ROOM
13 ELECTRIC ROOM
14 CEO'S ROOM

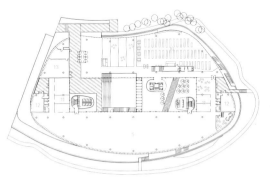

1ST FLOOR PLAN

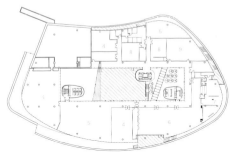

B1 FLOOR PLAN

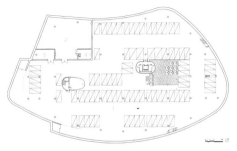

B2 FLOOR PLAN

Inner staircase leading to the underground floor.

The 1st floor lobby with an information desk

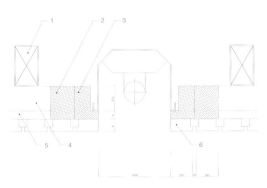
CEILING LIGHTING FIXTURE SECTION

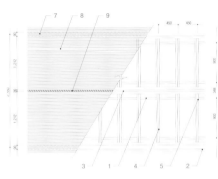
LOBBY CEILING PLAN DETAIL

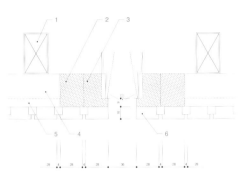
CEILING LINE DIFFUSER SECTION

1. 30X50 ST'L PIPE@900
2. 30X40 LAUAN MDF
3. 30X40 MDF MOULDING(ONE SIDE GROOVE CHIPPING)
4. 30X30 ST'L PIPE@450
5. 30X40 LAUAN MDF@450
6. T=16MM K-TOP LINIER 28/4 / FIN: LG DECO MW201Q3
7. LIGHTING FIXTURE
8. PET25T + K-TOP LINIER 16T
9. LINE DIFFUSER
10. THK16 K-TOP LINIER / TYPE: 28/4M / CORE MATERIAL: T=15MM MDF / FINISH: LG DECO MW201Q3
11. EDGE: PAINTING(FIELD)
12. 30X30 ST'L PIPE @900~1200

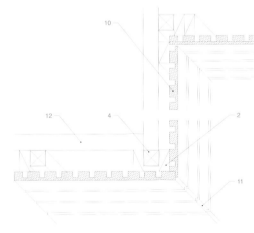
LOBBY CEILING CROSS SECTION

Office on the 2nd floor

1 SPRINKLER HEAD
2 ST'PLATE
3 LIGHTING FIXTURE
4 T20 FIBERGLASS CEILING MATERIAL
5 LINE TYPE DIFFUSER
6 SPEAKER
7 HANGER BOLT(@900-1200)
8 CARRING CHANNEL(@900-1200)
9 5.5X25LG PAN HD / SUS SCREW @350
10 FILED WEATHER SILICONE SEALANT / W / BACK UP ROD(BY OTHER'S)
11 GLASS SETTING CLIP
12 AL. EXTRUDE CLIP @350
13 EXPANTION JOINT
14 T28 / PAIR GLASS(8+12+8)(BY OTHER'S)
15 STRUCTURAL SILICONE SEALANT / WITH NORTON TAPE
16 E.P.D.M INNER / GLAZING GASKET
17 T1.2 E.G.I ST'L PLATE
18 FIRE STOP SPRAY(BY OTHER'S)
19 T100, 100K MINERAL WOOL(BY OTHER'S)
20 L-120X120X200LGX6T / FINISH: ANTI RUST PAINT / 2-COATED
21 ARC WELDING ON SITE
22 ST'L FASTNER 100X120X6T / FINISH: ANTI RUST PAINT / 2-COATED
23 T6 ST'L STOPPER
24 T2 AL. EXTRUSION COVER
25 L-20X20X2T 30LG. AL. ANGLE
26 CEILING LINE

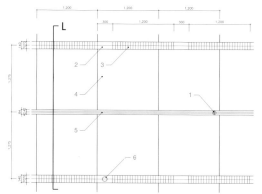

OFFICE CEILING PLAN

OFFICE CEILING CROSS SECTION L

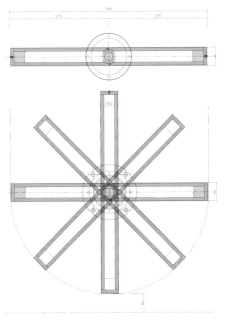

WOOD LOUVER PLAN

Office on the 2nd floor. The louvers that functionally control the awning are evocative of books.

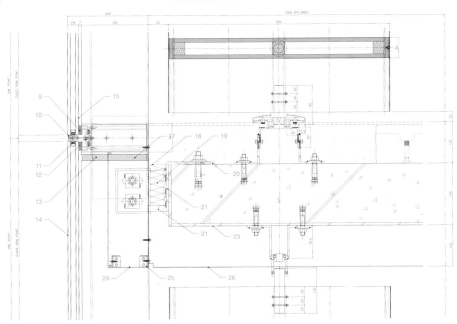

WOOD LOUVER VERTICAL SECTION

\>\> OFFICE

DMC OFFICE BUILDING (TRUTEC)

Location Sangam-dong, Mapo-gu, Seoul, Korea Site Area 2,078m² Building Area 1,184m² Total Floor Area 19,816.49m² Building Coverage Ratio 57.01% Floor Area Ratio 584.37% Building Scope B5–12F Height 54m Structure RC(Underground), RC & Steel(Ground) Exterior Finish Aluminum Curtain Wall, T24 Low-E Pair Pastel Glass, Zinc(Anthra Balck) Design Finish 2005.10 Construction Period 2005.7–2006.12 Architecture Design Frank Barkow, Regine Leibinger | Barkow Leibinger Architects Design Team Martina Bauer, Matthias Graf von Ballestrem, Michael Schmidt, Elke Sparmann, Jan-Oliver Kunze Structural Engineer Schlaich Bergermann and Partner, Jeon and Lee Partners Facade Engineer ARUP Facade Engineering Facade Company ALUTEK Contact Architecture Design Kim Byung-hyun | CHANGJO Architects Construction DONGBU Corporation Client Trumpf Korea Sang-Am Ltd. Photographer Lee Ki-hwan

The sure footing that we have enjoyed in the European building culture is giving way to new challenges in places like Asia where site, history, or methods may be irrelevant, transitional or appropriated from other places. The question of site-ed-ness is in itself a work-in-progress. Having relied on site as a key condition to which an architecture might react dialectically we have found ourselves in a position where site may not so much inform an architecture as simply receive one. In this sense our building for the Digital Media City is unabashedly self-referential. With neighboring buildings yet unplanned or unknown we found ourselves with no local context to respond to other than local zoning ordinances. Located in this new high-tech work-live center, one of the last remaining open sites in Seoul, the project is 11 stories of office and showroom space over a 5 level underground parking structure. The exterior cladding is a mirrored fractal glass articulated into a series of crystalline-formed bays projecting 50cm. These three-dimensional forms refract light and image, rendering the facade as a fragmented and abstract surface. Any less-than-fortunate examples of architecture surrounding it will be reduced to fragmented pixels upon its surfaces. Reciprocally from the interior, window bays act as kaleidoscopes offering multiple views and orientations through the facades. This way the building acts as a phenomenal filter, effecting both vision and how it is viewed. The other significant decision is to offset the building core to the probable lightless eastern corner of the building. The core is clad in a polygonal-shaped zinc shingle. This location of the core allows for a large representational building lobby to occur at the ground level with a mezzanine coffee shop above. A large triangular indentation marks the formal entrance to the building. This offset entrance also sets up a strong relationship to the park located diagonally across. A triangular suspended steel stair to the mezzanine flanks this entrance. Sectionally the building expands toward the lower levels, with higher spaces given for the ground floor (6m clear) and 1st and 2nd floor showrooms (4.5m). Text by Barkow Leibinger Architects

General view from the north-west

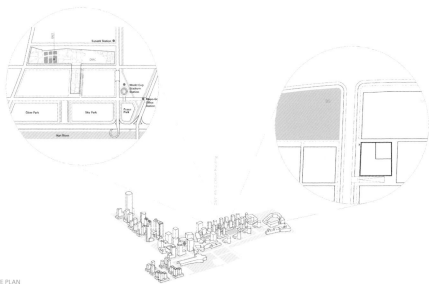

OFFICE
DMC OFFICE BUILDING(TRIUTEC)
SITE PLAN

ENTRANCE HALL & LOBBY STAIR PLAN

The main entrance where a huge triangular glass creates a prism.

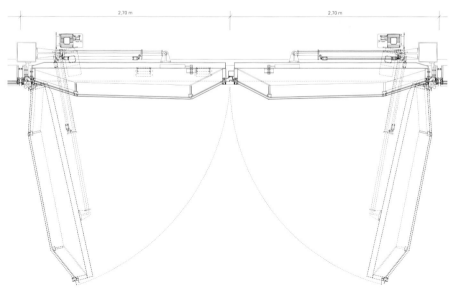

ENTRANCE DOOR DETAIL A

General view from the west. The 3-dimensional glass doors were installed considering the showrooms on the 3rd and 4th floors. The doors open and close by the mechanical automated system.

The 3-dimenional elevation reflects the light and images to create an abstract surface.

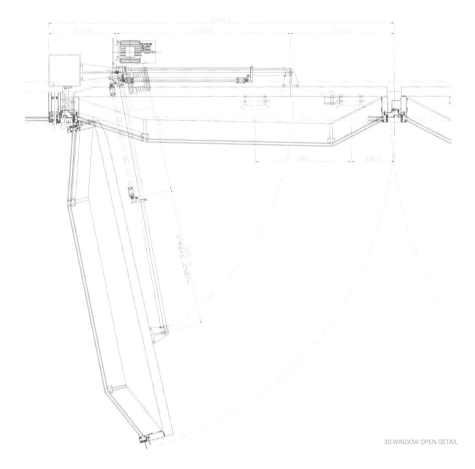

3D WINDOW OPEN DETAIL

3D TYPE CURTAIN WALL

As the first 3D type ever introduced in South Korea, the aluminum curtain wall system applied to TRUTEC building requires separate equipments for the 3D process of aluminum bars and the intensified maintenance for designing and mock-up process related to glass furring. Thus 5 axis process device exclusive for 3D was introduced to complete the curtain wall with three dimensional elevations.

For precision, the design mock-up was made before processing. The process involves examining the construction feasibility and finish on the connection area of 3D construction material that can be a problem when manufacturing the units. In modeling, Inventor, which is one of the 3D machine design programs, was used in order to decide the location of the processing device and clamps, the rotation speed of the processing device, the motion speed and the sequence of works. The aluminum bar processed through the simulation has various 3D connection areas that demand repeated tests on several functions unlike the case of existing 2D surfaces.

The curtain walls in the northeast, northwest and some southwest zones are three dimensional and the remaining zones are two dimensional. The 2D zones were planned according to the schedule of the temporary master planning and finish work. Taking into consideration the size, form and weight of the unit and the degree of difficulty of using temporary construction equipments, the constructor used a tower crane instead of a monorail and changed to an oil pressure crane after the dissection of the tower crane.

Text by Hong Du-pyo(DONGBU Corporation Co., Ltd.)

DESIGN MOCK-UP

FRAME MODELING & SPIGOT SAMPLE

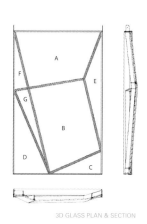

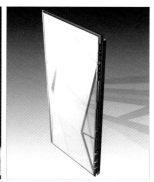

3D GLASS PLAN & SECTION

GLASS MODULE

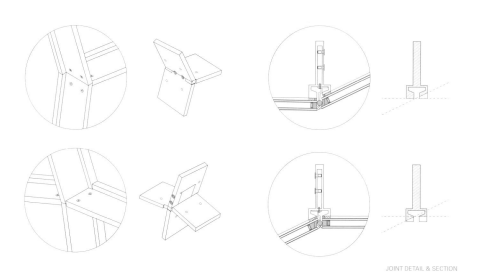

JOINT DETAIL & SECTION

The interior of the 7th floor office. It produces the sense of direction and various scenes as the patterns seen through a kaleidoscope.

1 FLOOR REGISTER WITH CONVECTOR AND UPLIGHT
2 STEEL COLUMN CLAD IN GALVANIZED METAL PANEL
3 STEEL BRACKET
4 PERIMETER FRAME
5 ANTIGLARE BLIND
6 FACADE MODULE FRAME
7 STEEL BEAM
8 CNC CUT ALUMINUM EXTRUSION
9 INTERMEDIATE MULLION
10 LOW-E INSULATING GLASS
11 SUSPENDED CEILING, GALVANIZED PERFORA-TED PANEL
12 ACCESS FLOOR
13 CONCRETE FLOOR ON CORRUGATED METAL
14 MAIN ENTRANCE
15 TYP. 3D GLAZING ELEMENT
16 TYP. 2D GLAZING ELEMENT
17 PARAPET
18 WINDOW WASHING SYSTEM

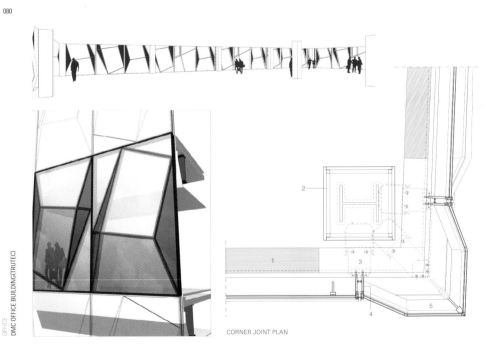

CORNER JOINT PLAN

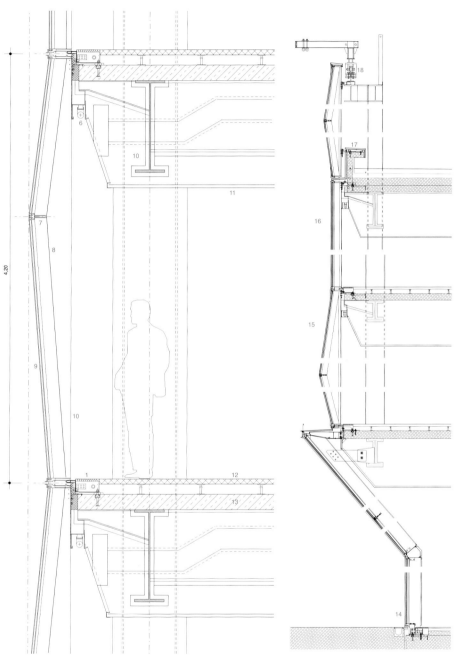

FACADE SECTION

FACADE CROSS SECTION

The steel triangular staircase. It is also related to the huge trigonal glass doors.

The inner staircase leading to the mezzanine floor.

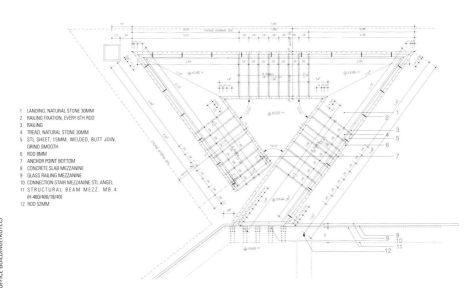

1 LANDING, NATURAL STONE 30MM
2 RAILING FIXATION, EVERY 6TH ROD
3 RAILING
4 TREAD, NATURAL STONE 30MM
5 STL SHEET, 15MM, WELDED, BUTT JOIN, GRIND SMOOTH
6 ROD 8MM
7 ANCHOR POINT BOTTOM
8 CONCRETE SLAB MEZZANINE
9 GLASS RAILING MEZZANINE
10 CONNECTION STAIR MEZZANINE STL ANGEL
11 STRUCTURAL BEAM MEZZ. MB.4 (H-460/400/18/40)
12 ROD 52MM

LOBBY STAIR SECTION

5~11TH FLOOR PLAN

1 ENTRANCE HALL
2 LOBBY
3 ELEVATOR HALL
4 APPLICATION CENTER
5 COFFEE SHOP
6 TRAINING & MEETING ROOM
7 OFFICE
8 PARKING LOT
9 SHOWROOM

MEZZANINE FLOOR PLAN

1ST FLOOR PLAN

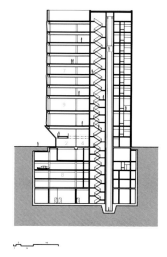

SECTION

\>\> OFFICE

DIGITAL MAGIC SPACE

Location Sangam-dong, Mapo-gu, Seoul, Korea Site Attribute Land Development, General Commerce Site Area 3,305.79m² Building Area 1,981.49m² Total Floor Area 16,341.19m² Landscape Area 1,308.53m² (39.6%) Building Coverage Ratio 59.94% Floor Area Ratio 327.83% Building Scope B2~12F Maximum Height 71.60m Structure SRC Design Period 2003.3~2003.12 Construction Period 2004.7~2006.3 Parking 129 Cars Architecture Design Jeong Young-kyoon, Park Gyu-yong | HEERIM Architects & Planners Construction KUMHO Engineering & Construction Client Korea Broadcasting Institute Photographer Lee Ki-hwan

Digital Magic Space (DMS) is not only the first symbol of the development of Digital Media Street (DMS) which is the core business of Sangam-dong Digital Media City (DMC), but it holds an even more significant image as it was selected as the location for global digital media exchange. For a problem-free supply of contents, in its early stage of development DMS focused on attracting media production industries and becoming the Mecca of high-tech media production centers by offering one-stop features including media production →exhibition →viewing, current affairs, events, public relations, consultations, and distribution. There were various requirements from stability to general functions even in its planning stage. From its conceptual design and over the four months of actual designing process, the plan intended to effectively satisfy all areas from external and internal functions, facilities, and neighboring environments. DMS is separated into three sections of the media production sector (lower floors 1~5 fl.), office and administration sector (upper floors 6~12 fl.), and external studio sector. These three sectors are linked via atrium lobby and roof-top terrace. To enhance the effectiveness of media production and related-facilities' work process, each floor was planned with pertinent features for its expected operation and consideration was taken for minimizing travel distance without congestion. On the top floor (12th fl.), a rest area and multi-purpose hall is in place to satisfy requirements such as movie production PR, movie premier, and others. The rest area is linked with the roof-top and has become one of the key factors in the building's upper section design and with spectacular lighting, it is designed to make DMS recognized as a landmark even from a distance. Overall, the sections from lower floors, mid floors, to upper floors, it's grandeur and layout is made harmoniously complete with the use of aluminum and glass.

1 MAIN ENTRANCE
2 SUB ENTRANCE
3 NEW BUILDING
4 PUBLIC OPEN SPACE
5 2ND PHASE WORK SITE

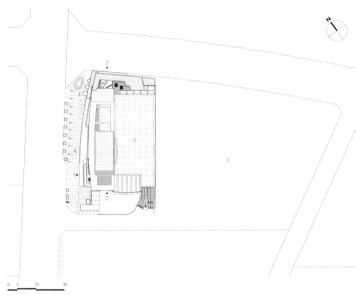

SITE PLAN

SOUTH ELEVATION　　　　NORTH ELEVATION　　　　EAST ELEVATION

The Digital Magic Space elevation is comprised of curtain wall.

1 STUDIO
2 OFFICE
3 STUDIO SUBCONTROL ROOM
4 EDITORIAL OFFICE
5 DUBBING STUDIO
6 STORAGE
7 SOUND STORAGE
8 SUBCONTROL ROOM
9 EXPERIENCE STUDIO
10 EDITORIAL PRODUCTION ROOM
11 DUBBING SUBCONTROL ROOM
12 CAMERA STORAGE
13 WAITING & DRESSING ROOM
14 PRACTICE ROOM
15 PARKING LOT
16 HVAC ROOM
17 WAITING & MEETING ROOM
18 DATA ROOM
19 LOBBY

CROSS SECTION

LONGITUDINAL SECTION

The sub entrance viewed from the lobby

Indoor of the front building

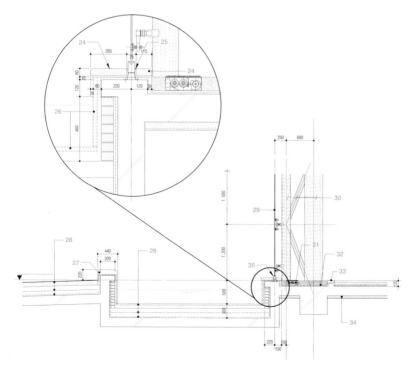

LOBBY FLOOR SECTION

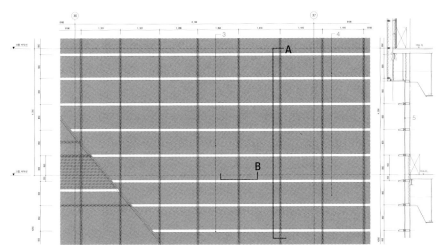

UPPER CURTAIN WALL ELEVATION & SECTION

1. T1.2 ST'L BACK PANEL (CORRUGATED STEEL PANEL)
2. T24 COLOR DOUBLE-FACED SEMI-TEMPERED PAIR GLASS
3. T24 COLOR LOW-E PAIRED GLASS (SINGLE-FACE TEMPERING)
4. T24 COLOR PAIRED GLASS (DOUBLE-FACED TEMPERING)
5. T24 COLOR PAIRED GLASS (DOUBLE-FACED SEMI-TEMPERING)
6. SILICON /NORTON TAPE
7. GASKET
8. T1.2 GL STEEL PANEL (CORRUGATED STEEL PANEL) / T75 GLASS WOOL (24K)/ T0.8 GL STEEL PANEL
9. MULTI LOCK HANDLE (2 LOCKING)
10. 4.5×16 LG SUS SCREW
11. ALUM EXPAND JOINT W/ALUM SLEEVE L =200
12. M12X110LG. HEX HD. BOLT & NUT/ W/ ST'L SQUARE WASHER
13. AL. WASHER (A6061-T6)
14. M16 × 60LG / COUPLING NUT
15. M16 × 60LG BOLT
16. ALUM. EXTRUDED SHEAR BLOCK
17. GLASS SETTING BLOCK Ø1/4 POINTS
18. WEATHER SILICONE SEALANT ON BACK-UP ROD SPONGE
19. THK6.0 STEEL WIND BRACER (ST'L GALV.)
20. 100 × 100 × 6T × 100L ST'L BENT PLATE(ST'L GALV.)
21. CAST-IN CHANNEL L=35
22. M16X60LG HIGH TENSION T-BOLT / NUT/ WASHER
23. 110 × 80 × 6T × 120LG / ST'L BENT PLATE(ST'L GALV.)
24. T50 GRANITE CAPPING STONE (RUBBING FIN.)
25. 10X10 CAULKING
26. GRANITE (LANDSCAPING WORK)
27. GRANITE INSTALL (LANDSCAPING WORK)
28. LANDSCAPING WORK/ T167 PLAIN CON'C W.M#8)/T3.0 SURFACE COATING WATERP-ROOFING
29. T12 TEMPERED GLASS
30. RED LEAD TWICE / OIL PAINT 3 TIMES
31. EMBEDDING TYPE CONVECTOR INSTALL
32. SOY BEAN GRAVEL PAVING LIQUID WATERPROOFING TYPE 1 / T20 PROTECTIVE MORTAR
33. T50 CAPPING STONE INSTALL
34. T50 RIGID HEAT INSULATION/ T10 SOUND-ABSORBING MATERIAL SPRAYING
35. T1.2 CLASS SHOE FLASHING

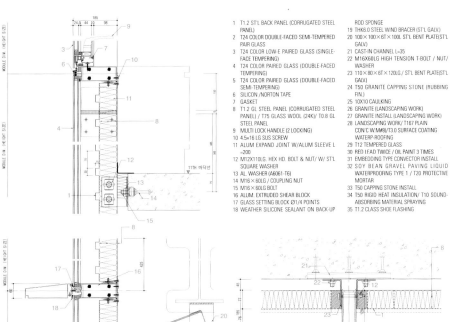

CURTAIN WALL SECTION A

CURTAIN WALL SECTION B

1. T12.5 GYPSUM BOARD 2 PLY / THK 90 METAL STUD(Ø400) / THK 50 GLASS WOOL 32K / T12.5 GYPSUM BOARD 2 PLY / THK 90 METAL STUD(Ø400) / THK 50 GLASS WOOL 32K / T12.5 GYPSUM BOARD 2 PLY / THK 90 METAL STUD(Ø400) / THK 50 GLASS WOOL 32K / T12.5 GYPSUM BOARD 2 PLY / THK 50 PVC JOINER(Ø1000X1000) / THK 50 G/W 48K/G / THK 50 CAP CLOCK(Ø330×330)
2. THK 90 METAL STUD(Ø400) / THK 50 GLASS WOOL 32K / T12.5 GYPSUM BOARD 2 PLY / THK 50 PVC JOINER(Ø1000×1000) / THK 50 G/W 48K/G / THK 50 CAP CLOCK(Ø330×330)
3. CURRENT WALL STRUCTURE / SQUARED LUMBER WOOD 45×45(Ø600×600) / THK50 GLASS WOOL 32K / THK9 PERFORATED MDF / FABRIC (OF SERIES)
4. WOOD BASE (H : 200)
5. MATERIAL SEPARATOR
6. CURRENT WALL STRUCTURE/ WALL DAMPER Ø 900/ SQUARED LUMBER WOOD 45×45(Ø450×450)/ T12.5 GYPSUM BOARD 2PLY/ SQUARED LUMBER WOOD 45X45(Ø450×450) - PART OF STRAIGHT LINE/ SQUARED LUMBER WOOD 45X45(Ø300) - PART OF CURVED LINE/ T4.8 2PLY /#20 WHITE COTTON CLOTH WALLPAPERING TWICE PUTTY ON THE WHOLE SURFACE/ APP. WATER-BASED PAINT
7. CEILING DAMPER Ø900/ SQUARED LUMBER WOOD 60×60(Ø900) / SQUARED LUMBER WOOD 45×45(Ø450×450) / T12.5 GYPSUM BOARD 2PLY
8. T3 DELUXE TILE/ T22 SELF-LEVELING/ T150 PLAIN CON'C (WIREMESH)/ T0.08 P.E. FILM 2PLY SPREADING/ T12 WATERPROOF PLYWOOD 2PLY/ T50 DAMPING PATH Ø900×900/ FLOOR SLAB
9. CURRENT WALL STRUCTURE / WALL DAMPER Ø900/ SQUARED LUMBER WOOD 60×60(Ø900)/ SQUARED LUMBER WOOD 45×45(Ø450×450)/ T12.5 GYPSUM BOARD 2PLY/ SQUARED LUMBER WOOD 45×45(Ø450×450)/ T50 GLASS WOOL 32K / FABRIC (OF SERIES)/Ø4 WIREMESH (Ø50×50) / POWDER COATING (APP. COLOR)/ T11×50 STEEL MOLDING (Ø900×1800) / POWDER COATING (APP. COLOR)
10. T4.0 AL COMPOSITE PANEL
11. Ø9 HANGER BOLT Ø900×900
12. CEILING DAMPER Ø900×900
13. GRID IRON(POSITION ADJUSTING · ACOUSTICS)
14. STEEL BASEBOARD (H : 200)
15. SOUND-ISOLATED TEST WINDOW
16. WATER-BASED PAINT

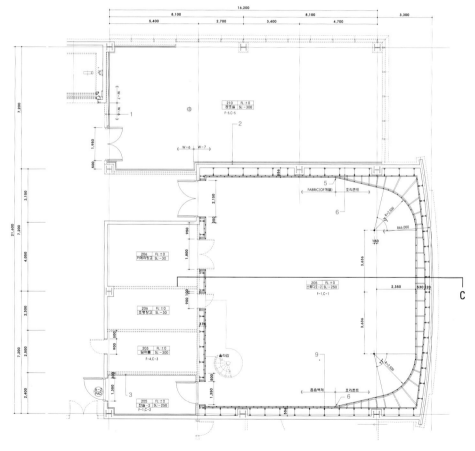

1ST FLOOR PARTIAL PLAN(STUDIO)

Open space of the studio

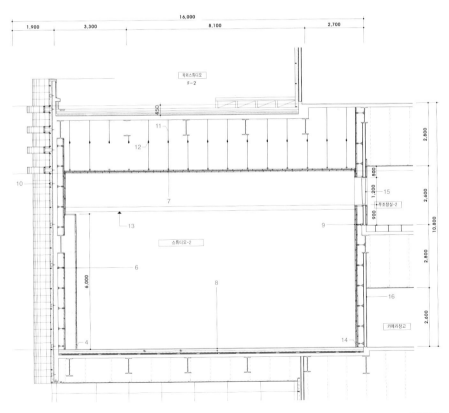

SECTION C

1. 1.6 GALVANIZED STEEL PANEL (PERFORATED PANEL) / AFTER READ LEAD TWICE/ OIL PAINT 3 TIMES/ POWDER COATING FIN.
2. BEGINNING LINE OF TOP STEP
3. T4.5 CHECKED STEEL PANEL (2ND BASEMENT/ ENGINE ROOM) OR T4.5 CHECKED STEEL PANEL / T24 CEMENT MORTAR/ T2 INORGANIC FLOOR REINFORCING AGENT (1ST, 2ND, 4TH FLOOR / STUDIO)
4. Ø50XT1.5 STAINLESS STEEL PIPE/ HAIRLINE FIN.
5. Ø216.3XT5.8 STEEL PIPE/ OIL PAINT 3 TIMES AFTER RED LEAD 2 TIMES
6. HANDRAIL POST/ 60XT12 STEEL F.B./ OIL PAINT 3 TIMES AFTER RED LEAD 2 TIMES
7. OIL PAINT 3 TIMES AFTER RED LEAD 2 TIMES AFTER T12 STEEL PANEL WELDING
8. MATERIAL SEPARATOR (MORTAR FILLING)
9. T9 STEEL PLATE
10. FIELD WELDING
11. RED LEAD TWICE / OIL PAINT 3 TIMES
12. Ø16 SET ANCHOR

STAIR PLAN & SECTION

DETAIL D

PERFORATED PANEL PLAN

STAIR PLAN

SECTION E

STAIRS SECTION(STUDIO)

STRUCTURE & FINISH PROCESS

STACKED-TYPE STUDIO AND CONSTRUCTION NOISE DETAIL The first challenge during the design was how to utilize a larger than expected area for the cooperative usage room. Upon contemplation, a grating floor method was selected for the cooperative usage room so that it will not be included in the calculation of floor area; and to meet the noise level standards required by neighboring studios, equipments and machines with spring dampers installed were designed. Since three studios are being designed over a comparable small and confined area with consideration to connections to relative offices from each studio, a stacked-type studio, which is rare in Korean architecture, was designed. The studios stacked from floors 1~3 and floors 4~6 had to meet the noise standards for each studio with the slab flooring of the 4th floor as the boundary. To resolve this issue, the slab flooring of the 4th floor was structured as a raised floor so that the noise and vibration from the 4th floor does not transfer to lower studios. This method was created through noise and vibration simulation under the notion that since the building will not be equipped with a separate stage production facility as large studios are, we considered a possibility that stage production may be conducted on the 4th floor. In addition to the aforementioned studios, other broadcasting facilities from the 1st to 6th floors includes a control room, anchor booth, dubbing room, make-up room, and other related facilities which are all designed to meet or exceed the standards used in actual broadcasting centers.

CHANGE IN CURTAIN WALL DESIGN
- Lower Floors: From the 1st floor to the 3rd floor, the SPG System, which clearly opens the front facade section of the lobby has the effect of making the structure look bigger from the outside. On the outer section of SPG, a louver is installed and to install a casement window, the upper section of the lobby was changed to a general curtain wall system which slightly reduced the curtain wall's weight.
- Typical Floors: There are always some parts which require a change or modification from the blue print during actual construction process. In the case of this project, the main change was that at the time of the design, the unit-type curtain wall system was specified but was changed to stick-type curtain wall. For the construction, analysis from various perspectives was considered and I have agreed to the change to stick-type but as the architect,

it still remains a regret. The multi-purpose aluminum panel used together with 24T low-e paired glass, was selected for the finish with three coats of free coating with consideration to discoloration, disconfiguration, and applicability.

MIXED SLAB STRUCTURE AND HEAT RESISTANCE ISSUE The structure of the building is designed with a steel frame, steel reinforced concrete beams with steel supports, tech slab above steel-reinforced concrete mixed with slab. Therefore, although it is highly effective in heat resistance required in the top floor's standards and its construction applicability, if non-steel reinforced concrete and other finished section cracks, applied Styrofoam will become wet, causing large problems of leakage and thus, in certain sections of the lower deck parts of the inner side of the building, heat-resisting material was selected as a secondary method.

1 CONTROL OFFICE
2 STORAGE
3 DIGITAL NETWORK PLAZA
4 MAIN ENTRANCE
5 SUB ENTRANCE
6 STUDIO
7 LIGHTING STORAGE
8 SOUND STORAGE
9 CAMERA STORAGE
10 ROOM
11 PRACTICE ROOM
12 SET STORAGE
13 DRESSING ROOM
14 HVAC ROOM
15 DIMMER ROOM
16 SUBCONTROL ROOM
17 MACHINERY ROOM
18 DUBBING STUDIO
19 DUBBING SUBCONTROL ROOM
20 OVERALL EDITORIAL OFFICE
21 DATA ROOM
22 EDITORIAL PRODUCTION ROOM
23 OUTDOOR STUDIO
24 WAITING & MEETING ROOM
25 BOOTH
26 EDITORIAL OFFICE
27 OFFICE

2ND FLOOR PLAN

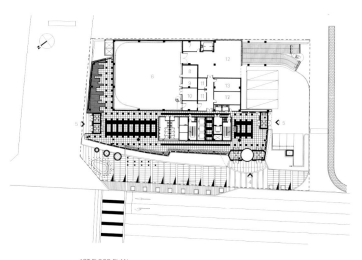
1ST FLOOR PLAN

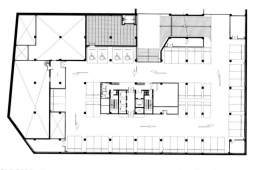
B1 FLOOR PLAN

4TH FLOOR PLAN

7TH FLOOR PLAN

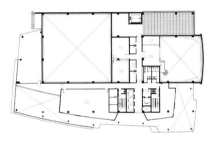

3RD FLOOR PLAN

5TH FLOOR PLAN

Outdoor garden on the 4th floor

Outdoor garden on the 7th floor

\>\> OFFICE

KORLOY HOLYSTAR BUILDING

Location Doksan-dong, Geumcheon-gu, Seoul, Korea Site Attribute General Resident, General Aesthete, Maximum Height Site Area 1,111.80m² Building Area 555.27m² Total Floor Area 4,446.89m² Landscape Area 169.07m² Building Coverage Ratio 49.94% Floor Area Ratio 249.19% Building Scope B2-6F Structure RC Exterior Finish Metal Fabric, Stainless Steel Panel, Macheon Stone Burner Flaring, T24 Clear Low-E Pair Glass Interior Finish Polishing Tile, Vinyl Paint on Gypsum Board Design Period 2005.5-2005.7 Construction Period 2005.12-2006.11 Architecture Design MANO Architecture Inc. Design Team Yeo Sang-kwon, Lee Jong-hwa, Park Hyun-woo, Lee Byeong-jae Client KORLOY HOLYSTAR Photographer Lee Ki-hwan

General view from the road

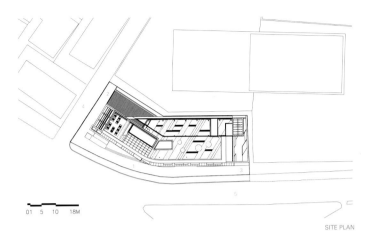

1 MAIN ENTRANCE
2 UNDERGROUND PARKING LOT ENTRANCE
3 SUB ENTRANCE
4 6M ROAD
5 10M ROAD

SITE PLAN

REAR ELEVATION

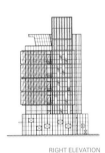

RIGHT ELEVATION

CREATING NEW BRAND CULTURE

SITE · PROGRAM Located at the intersection of southern highway and Siheung interchange, the site is a very narrow and L-shaped land. There is 3m setback line in front and a 6m road with 1m level difference on the left. Except for officetel in the back, old neighborhood facilities and residential districts surround the site. The primary request of the client was to design a high-tech and future-oriented corporate image to represent international ultra lightweight hardware manufacture company. Secondly, maximum floor area was essential for rental offices on the first to the third floors, and corporate offices on the fourth to sixth floors. In addition, enough parking spaces are provided for visitors' conveniences.

BRAND CULTURE Based on the given requests, core and office spaces are optimized to increase maximum feasibility and economical efficiency. Centered on the core system, vehicle circulation was efficiently designed with skip floors. To keep up with the international brand image, the building accommodates various community spaces such as meeting spaces, an atrium, roof gardens, exterior stairs with terraces, and pocket gardens on the first floor, etc. These kinds elements were architectural hardwares adopted to raise its own brand culture.

ARCHITECTURAL DESIGN Extruding windows are finished with metal fabrics and stainless steel panels to strengthen its sculptural massiveness and to give high-tech and global corporate image. In the narrow and curvy shape of the site, the main issues were how to design core and office zones efficiently, while securing enough parking spaces. Compact core system was allocated on the left with main utility facilities such as rest rooms and access holes, and skip floor system maximizes the number of parking lots. To emphasize horizontally long elevation, the building was divided into three parts; a base(the first floor), a joint(the second floor), a body(the third to the sixth floors), and finally a rooftop floor. Additionally, different kinds of sculptural elements and exterior materials were applied to each part.

Exterior designed with pop-up window

1. T1.2 SST'L HONEYCOMB PANEL
2. METAL FABRIC (MELAPIA 2404)
3. T1.2 SST'L FLASHING
4. T24 LOW-E PAIRED GLASS
5. T6 TEMPERED GLASS
6. HANDRAIL (TYP.)
7. T3 AL. SHEET / FLUOROCARBON RESIN COATING(BLACK)
8. T30 GRANITE GRINDING PO-CHEON STONE
9. T30 C-BLACK GRINDING
10. ATMOSPHERIC CORROSION RESISTANT STEEL PLATE PLANT BOX
11. MANO ARCHITECTURE INC(LETTER SHAPE- INTAGLIO, FONT- BANK GOTHIC MEDIUM BT, LETTER HEIGHT-6CM, SPACE-1CM)
12. C-BLACK GRINDING
13. BOXWOOD TREE GREGARIOUSLY PLANTING (UPPER FACE ARRANGEMENT)
14. T12 CLEAR TEMPERED GLASS
15. ATMOSPHERIC CORROSION RESISTANT STEEL PLATE PLANT BOX(TYP.)
16. T1.6 SST'L FLASHING
17. SMOKE WINDOW
18. T50 C-BLACK GRINDING (CAPPING STONE)
19. T3 AL. SHEET / FLUOROCARBON RESIN COATING
20. STRUCTURAL CAULKING
21. T12 CLEAR TEMPERED GLASS
22. APP. URETHANE PAINT / T50 MORTAR
23. T1.2 ST'L PLATE / POWDER COATING
24. W=200 GALVANIZED TRENCH
25. REFER TO FINISHES BY ROOF TYPE / T150 PLAIN CON'C(#8-150X150 WIRE MESH)
26. INTERIOR FIN.
27. INTERIOR FIN. / T24 PROTECTIVE MORTAR
28. INTERIOR FIN. / O.A FLOOR
29. T6 HIGH-PERFORMANCE INSULATION MATERIAL / T1.2 SST'L FLASHING
30. ASBESTOS-FREE SOUND-ABSORBING TEX
31. MORTAR / VINYL TILE
32. T50 PRESSURE-METHOD INSULATING BOARD (SPECIFIC GRAVITY 0.5 OR UNDER)
33. APP. URETHANE PAINT / T100 PLAIN CON'C (#8-150X150 WIRE MESH) / T24 PROTECTIVE MORTAR / T3 MEMBRANE WATERPROOFING
34. BUILDING LIMIT LINE
35. BOUNDARY LINE OF ROAD
36. T3 POLYCARBONATE
37. T30 C-BLACK GRINDING / T50 MORTAR / T100 PLAIN CON'C / T150 CODDLE STONE / GROUND COMPACTION
38. T100 PLAIN CON'C / T24 PROTECTIVE MORTAR / T3 MEMBRANE WATERPROOFING
39. 150X150 GRANITE BOUNDARY STONE
40. ARTIFICIAL SOIL INFILLING
41. BOXWOOD TREE GREGARIOUSLY PLANTING
42. T6 ATMOSPHERIC CORROSION RESISTANT STEEL PLATE PLANT BOX
43. Ø50 P.V.C DRAIN PIPE @3000(T.Y.P)

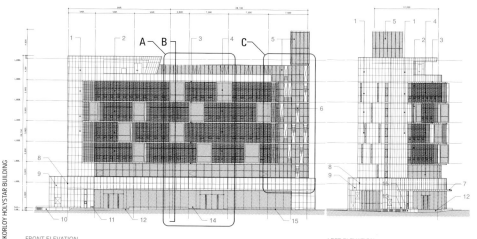

FRONT ELEVATION

LEFT ELEVATION

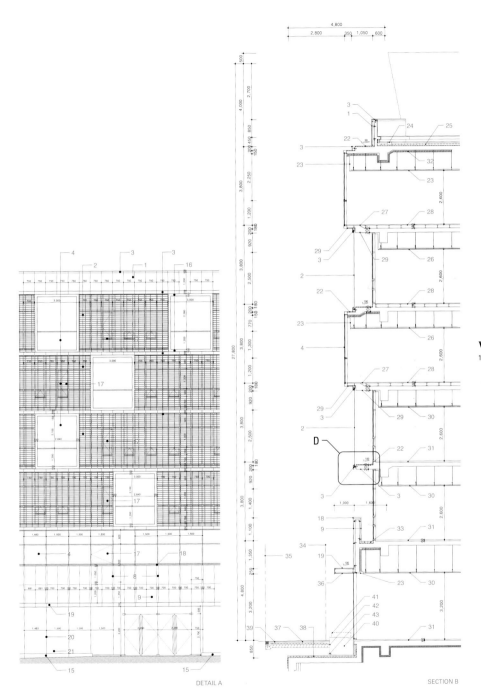

DETAIL A   SECTION B

The pop-up window viewed form the outdoor

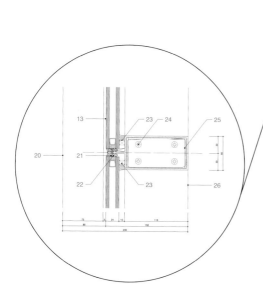
CURTAIN WALL JOINT DETAIL

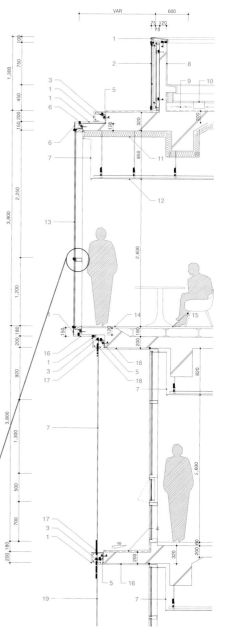
POP-UP WINDOW SECTION

1 T1.2 SST'L FLASHING
2 T1.2 SST'L HONEYCOMB PANEL
3 PRODUCT-FIXING HARDWARE 142X10T CONT'
4 APP. URETHANE PAINT / T50 MORTAR
5 L-(80+100)X150X6T
6 CAULKING
7 T1.2 ST'L PLATE / URETHANE PAINT
8 T18 MORTAR
9 REFER TO FINISHING ACCORDING TO ROOF TYPE / T150 PLAIN CON'C(SLPOE: 1/1000) (#8-150X150 WIRE MESH) / T24 PROTECTIVE MORTAR / T3 MEMBRANE WATERPROOFING
10 W=200 GALVANIZED TRENCH
11 T150 PRESSURE-METHOD INSULATING BOARD (SPECIFIC GRAVITY 0.5 OR UNDER)
12 LIGHT GAUGE STEEL CEILING RIB / INTERIOR FIN.
13 T24 LOW-E PAIRED GLASS
14 INTERIOR FIN. / T24 MORTAR
15 INTERIOR FIN. / O.A FLOOR
16 T6 HIGH-PERFORMANCE INSULATION MATERIAL / T1.2 SST'L FLASHING
17 Ø 6 CAP BOLT & NUT
18 Ø10 SST'L CIRCULAR BAR
19 METAL FABRIC(MELAPIA 2404)
20 MULLION CAP EDGE LINE
21 BACK UP
22 SEALANT
23 STRUCTURAL GLAZING SILICON SEALANT WITH NORTON TAPES
24 6DIAX20LG, 4NOS PER EACH CLIP
25 AL. SHEAR BLOCK ⊏ -70X107X70X3T-74LG
26 NULLION EDGE LINE
27 Ø12 SET-ANCHOR
28 THK.10 SST. COMPOSITE PANEL
29 MELAPIA
30 Ø12 CAP BOLT & NUT
31 PRODUCT-FIXING HARDWARE 80X100X10T (@200-400)
32 FLAT BAR 60XCONT'X10T

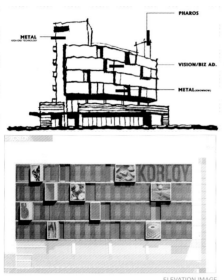

ELEVATION IMAGE

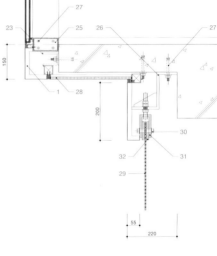

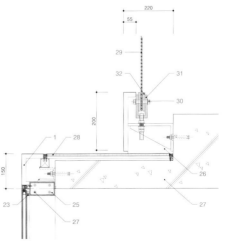

The pop-up window on the exterior elevation has been made of high-tech materials such as metal fabric and stainless steel panels. Used as a lounge for relaxation and meeting at normal times, it also functions as an advertising bulletin board for the company when put up with advertisement. The metal fabric, which originates from Korean traditional screen 'bal', is applied to the projected slab platform and creates double skins along with the inner curtain wall, thus connecting and harmonizing the interior and exterior spaces in a natural fashion. And it is a metaphorical representation of the knowhow preservation and the unique business area of the company.

METAL FABRIC SECTION

The void staircase is connected to the outside.

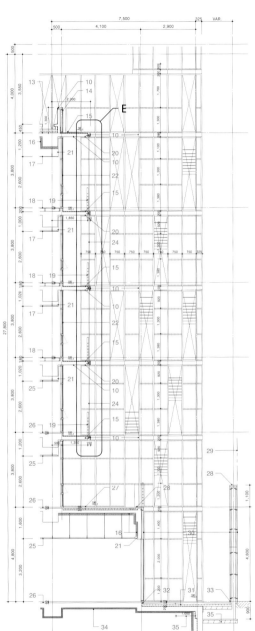

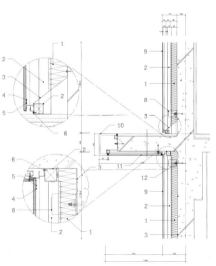

DETAIL D

SECTION C

1. T75-24K GLASS WOOL
2. □-50X50-2.3T
3. THK.0.5 E.G.I
4. SEALANT
5. RIVET
6. Ø4-16 BITS
7. CONCRETE FIN
8. WELDING
9. THK.10 SST. COMPOSITE PANEL / TOP-THK.1.0 STS 316 BEAD BLAST / MIDDLE-THK.0.4 STS 304 / BOTTOM-THK.0.4 STS 304
10. T1.2 SST'L FLASHING
11. THK.10 SST. COMPOSITE PANEL
12. FASTENER / L-(100+80)X80-6T / Ø12-100 SET ANCHOR
13. REFER TO FINISHING ACCORDING TO ROOF TYPE / T150 PLAIN CON'C (#8-150X150 W-MESH) / T24 PROTECTIVE MORTAR / T3 MEMBRANE WATERPROOFING
14. T1.2 SST'L HONEYCOMB PANEL
15. APP. URETHANE PAINT / T50 MORTAR
16. T150 PRESSURE-METHOD INSULATING BOARD (SPECIFIC GRAVITY 0.5 OR UNDER)
17. INTERIOR FIN.
18. INTERIOR FIN. / O.A. FLOOR
19. HANDRAIL
20. WATERDROP GROOVE (15X15)
21. T1.2 ST'L PLATE / POWDER COATING
22. T24 LOW-E PAIRED GLASS
23. T6 TEMPERED GLASS
24. METAL FABRIC(MELAPIA 2404)
25. ASBESTOS-FREE SOUND-ABSORBING TEX
26. MORTAR / VINYL TILE
27. APP. URETHANE PAINT / EXPANSION JOINT /T100 PLAIN CON'C (SLOPE : 1/1000)(#8-150X150 WIRE MESH) / T24 PROTECTIVE MORTAR / T3 MEMBRANE WATERPROOFING
28. T50 C-BLACK GRINDING (CAPPING STONE)
29. BOUNDARY LINE OF ADJACENT SITE
30. T30 C-BLACK GRINDING / T30 PO-CHEON STONE GRINDING
31. T30 C-BLACK GRINDING / T30 PO-CHEON STONE GRINDING / T50 MORTAR / T100 PLAIN CON'C (#8-150X150 WIRE MESH) / T24 PROTECTIVE MORTAR / T3 MEMB-RANE WATERPROOFING
32. T30 C-BLACK GRINDING
33. WATERPROOFING MORTAR / Ø10 P.V.C WATER-PIPE @3000
34. T100 ROCK WOOL SPRAYING
35. WATERPROOF LIMITS
36. T60 LEVELING CON'C / T150 CODDLE STONE
37. Ø12 SET-ANCHOR
38. T-(200+200)XCONT'X10T
39. MELAPIA
40. Ø12 CAP BOLT & NUT
41. PRODUCT-FIXING HARDWARE 80X100X10T (@200-400)
42. FLAT BAR 60XCONT'X10T

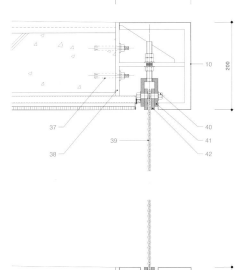

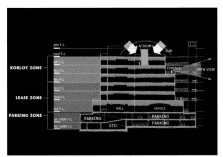

SECTION

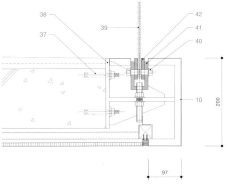

DETAIL E

The atrium on the 6th floor

The landscape viewed form the office on the 1st floor

1 UNDERGROUND PARKING LOT	6 HALL	ROOM
	7 WATER TANK	13 OFFICER'S ROOM
2 STORAGE	8 MAIN ENTRANCE	14 CEO'S ROOM
3 ELECTRIC ROOM	9 CONTROL OFFICE	15 ATRIUM
4 GENERATOR ROOM	10 OFFICE	16 ROOF GARDEN
	11 TERRACE	17 RESTING ROOM
5 MACHINE ROOM	12 CONFERENCE	

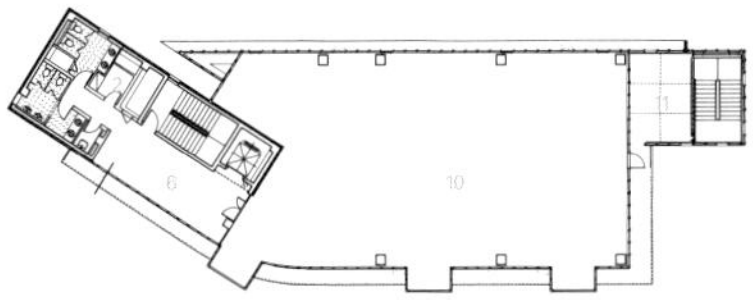

3RD FLOOR PLAN

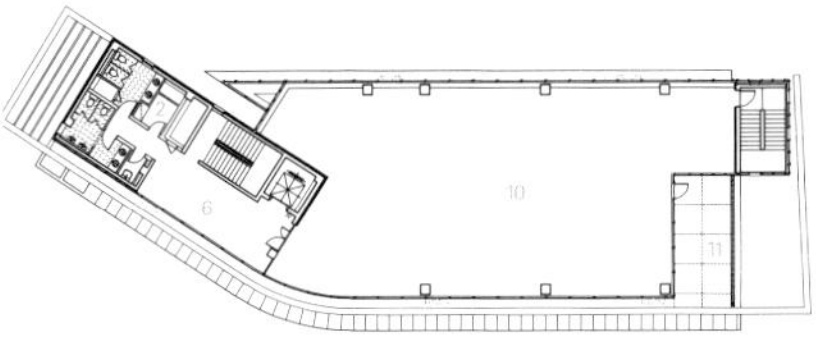

ROOF FLOOR PLAN

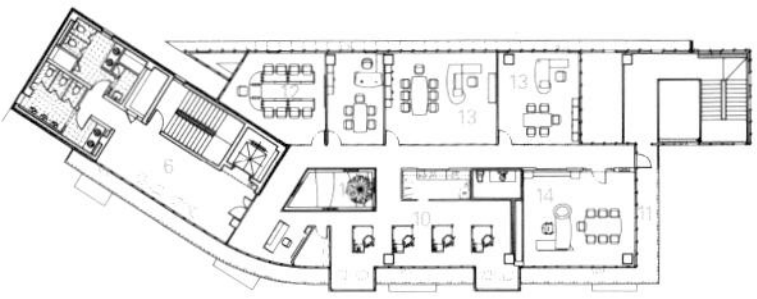

2ND FLOOR PLAN

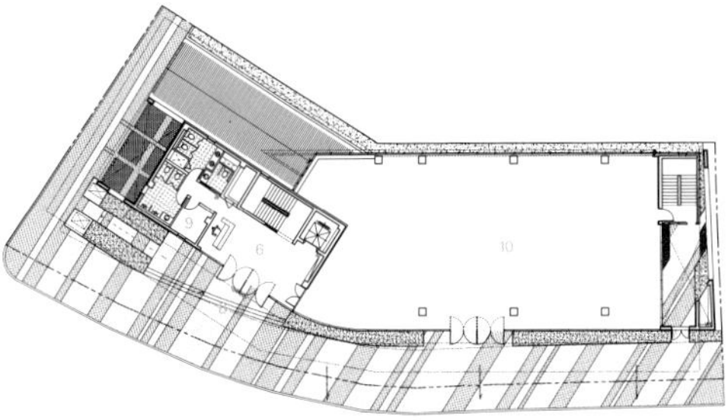

6TH FLOOR PLAN

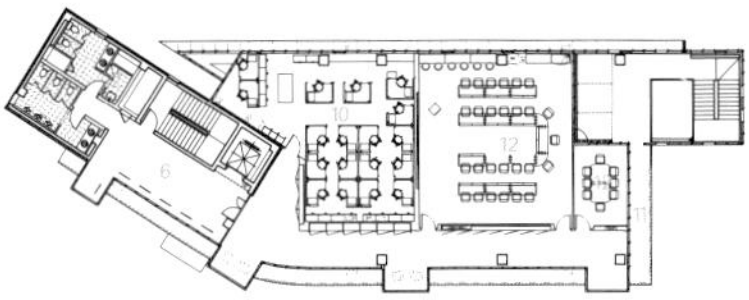

1ST FLOOR PLAN

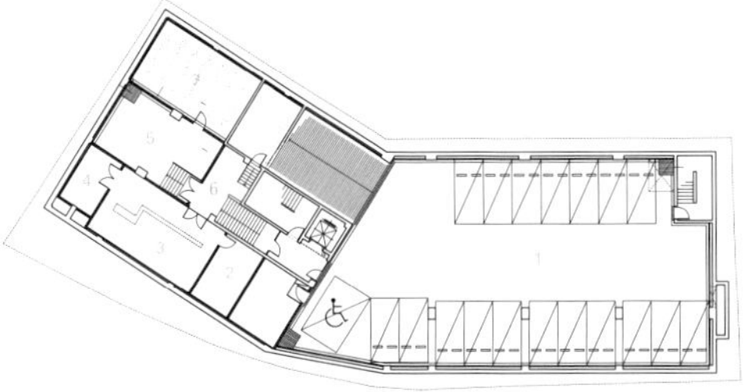

5TH FLOOR PLAN

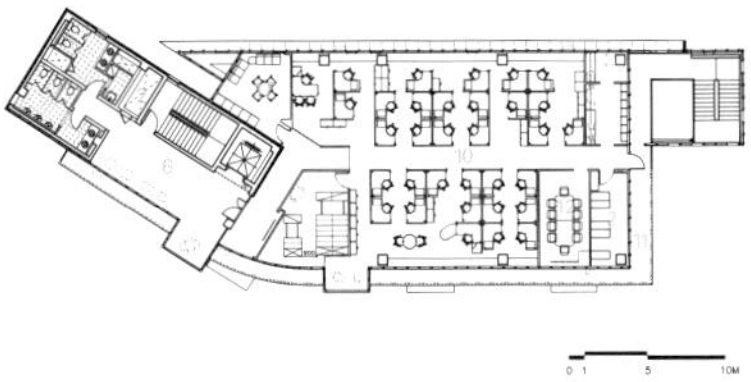

B2 FLOOR PLAN

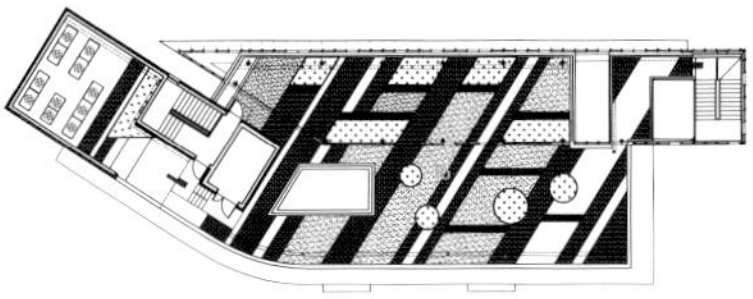

4TH FLOOR PLAN

\>\> OFFICE

B2Y OFFICE

Location Dongcheon-dong, Suji-gu, Yongin-si, Gyeonggi-do, Korea Built Area 1,320m² Completion 2008.11 Design Yoon Seok-min | Y SPACE Design Team Kim Min-jung, Son Lyang-kyung, Kim Hyoun-kwan Client B2y Photographer Kim Young, Yun Jae-hyun

Lines can create images by themselves. They can also create surfaces and canvases when an infinite number of them are gathered together. Using linear materials and harnessing the freedom of lines to its full capacity, Principal Yoon Sukmin, who studied painting, has rendered the lock of a girl's hair and created a canvas with extraordinary texture that is soaked in scarlet ambience. Compared to its status as the world's number one exporter in beauty business, B2y is still unknown to many people in its home country of Korea. With a view to strengthening their corporate image through remodelling of its headquarter office, B2y has chosen an extraordinary design that uses 'cable ties' in a competition. With the shape of countless lines flowing from the facade, a single image of hair overlaps with the headquarters of the manufacturer of hairdressing equipments. The remodelling project for the interior spaces consists of the public areas and the functional spaces in the fourth floor. Inside the entrance, the white staircase leading to the fourth floor at one stretch and the large lighting drops hanging from the ceiling create a sense of dynamic movement. On the fourth floor, the top floor, visitors are once again greeted by 'cable ties' which form the ceiling this time. Scarlet light seeps through the white hairs, creating a dreamy atmosphere. The space leads to the instruction room, where mirrors with built-in lamps reflect each other to create dynamic images. On the other side of the narrow hallway, which functions also as showcases for B2y's products, is another lounge, through which the conference room and the visitors' lounge are accessible. On the interior walls of the headquarters, white finishing materials with different heights alternate against the black background, creating an image of flowing lines, inspiring a sense of stability and change at the same time.

Main entrance

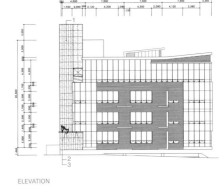
ELEVATION

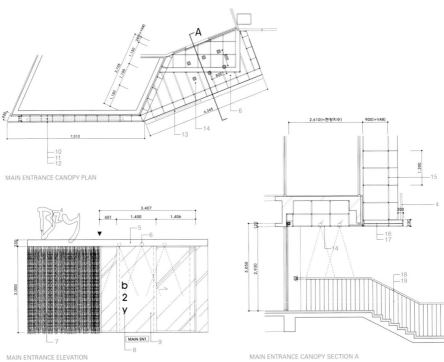

MAIN ENTRANCE CANOPY PLAN

MAIN ENTRANCE ELEVATION

MAIN ENTRANCE CANOPY SECTION A

Lobby on the 1st floor

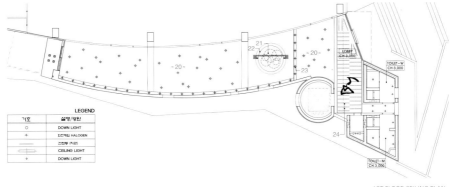

1ST FLOOR CEILING PLAN

기호	설명/광원
○	DOWN LIGHT
+	진열색시 HALOGEN
▬	간접등 (직관)
	CEILING LIGHT
○	DOWN LIGHT

LEGEND

1 APP' WIRE NET / WHITE CABLE TIE(INSERT LIGHTING-INSIDE PROJECTOR)
2 SIGN-NEON
3 CANOPY: T1.6 GALVA / BLACK METAL PA-INT
4 SIGN LOGO SCASI / INSIDE-NEON LIGHTING
5 CANOPY: BLACK PAINT FIN.
6 HALOGEN DOWN LIGHTING(FOR OUTSIDE)
7 APP' WIRE NET / CABLE TIE
8 SIGN: T5 ACRYLIC FIN (INSERT LIGHTING)
9 AUTO DOOR: BLACK PAINT FIN.
10 SQUARED PIPE REINFORCEMENT
11 INSIDE SIGN-NEON
12 APP' WIRE NET / WHITE CABLE TIE(INSERT LIGHTING-INSIDE PROJECTOR)
13 OUTSIDE SIGN-NEON
14 LOWER PART CANOPY : SQUARED PIPE REINFORCEMENT / T1.6 GALVA / BLACK METAL PAINT
15 SQUARED PIPE REINFORCEMENT / APP' WIRE NET / WHITE CABLE TIE(INSERT LIGHTING-INSIDE PROJECTOR)
16 T9 STEEL PLATE(SLOPE)
17 BLACK BAKING PAINTING / T1.6 GALVA / BLACK PAINT FIN.
18 HANDRAIL : T8 WHITE ACRYLIC / INSERT DECO LAMP
19 EXISTING FLAT STEEL / BLACK PAINT FIN.
20 APP' V.P FIN.
21 MAKING LIGHTING : 5EA
22 Ø=2,000
23 4TH FLOOR WALL : UP LIGHTING
24 APP' RED NEON / APP' BARRISOL FIN. (INSERT LIGHTING)

Main stair with pendent lighting of the hall

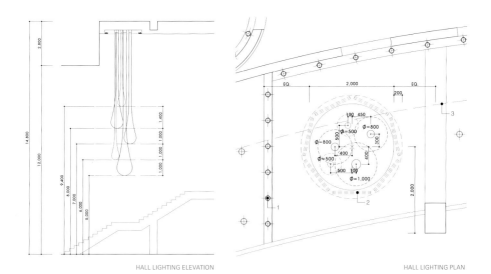

HALL LIGHTING ELEVATION

HALL LIGHTING PLAN

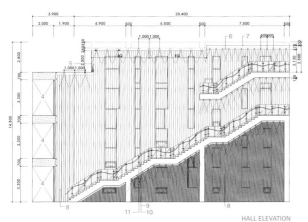

1 4TH FLOOR WALL UP LIGHTING
2 FL W=900
3 LOWER STAIR LINE
4 OPEN
5 APP UP LIGHTING
6 BEADS CHAIN, MAKING LIGHT BOX / FRONT: WHITE PAINT FIN.
7 NEW CEILING: WHITE PAINT FIN.
8 APP. PAINT FIN. / BEADS CHAIN CURTAIN WALL
9 HAND RAIL: T10 TEMPERED GLASS
10 HAND RAIL: Ø40 PIPE SST'L H/L
11 T9 STEEL PLATE
12 BACK: FROST SHEET 1PLY / FRONT: LINE FROST SHEET, BLACK LINE SHEET 1PLY

HALL ELEVATION

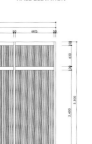

HALL PARTIAL ELEVATION

DETAIL B

Cable tie and red color lighting make unusual ceiling in lounge

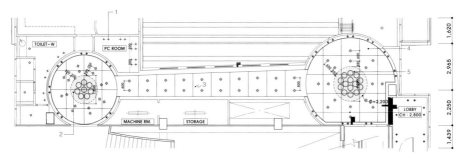
4TH FLOOR HALL CEILING PLAN

The lobby on the 4th floor | Narrow corridor with showcases

1 LIGHTING L5 - SPOT HALOGEN / Ø95 PERFORATING
2 APP. WIRE NET / H:280, WHITE CABLE TIE('RED COLOR' RING SHAPED FLUORESCENT LAMP 8EA)
3 APP. WIRE NET / WHITE CABLE TIE
4 APP. WIRE NET / H:140, WHITE CABLE TIE
5 APP. WIRE NET / H:280, WHITE CABLE TIE('RED COLOR' RING SHAPED FLUORESCENT LAMP 14EA)
6 INSERT LIGHTING
7 SHOWCASE: T10 CLEAR GLASS FIN.
8 WHITE PAINT FIN.
9 GROOVE: BLACK PAINT FIN.

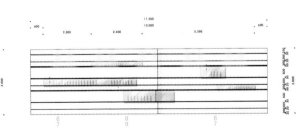

4TH FLOOR HALL CORRIDOR ELEVATION

Black&white lined conference room

Practice teaching room

Toilet

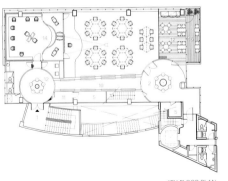

4TH FLOOR PLAN

1 ENTRANCE
2 LOBBY
3 HALL
4 INFO DESK
5 STORAGE
6 OFFICE
7 CONFERENCE ROOM
8 SNACK ROOM
9 MEETING ROOM
10 CORRIDOR
11 MACHINE ROOM
12 LOUNGE
13 LAUNDRY ROOM
14 BEAUTY SALON
15 TERRACE

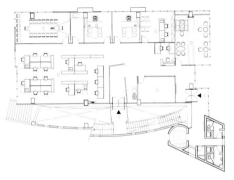

2ND FLOOR PLAN

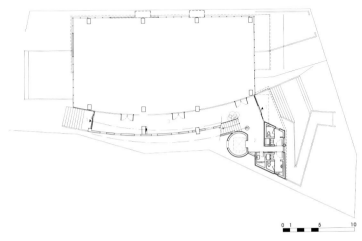

1ST FLOOR PLAN

>> OFFICE

JEOLLANAM-DO PROVINCIAL GOVERNMENT OFFICE

Location Namak-ri, Samhyang-myeon, Muan-gun, Jeollanam-do, Korea Site Area 231,781m² Building Area 20,827m² Landscape Area 107,180m² Total Floor Area 79,305.3m² Building Coverage Ratio 8.99% Floor Area Ratio 27.98% Maximum Height 106.06m Building Scope B2–23F Structure SRC, Steel Truss, Steel Exterior Finish T30 Granite Smooth Trimming, Galvanized Fluorocarbon Resin-Steel Plate, Extruded Cement Panel, Paint, Aluminum Corrugated Board, Aluminum Curtain Wall, T24 Clear Pair Glass Interior Finish Granite (Rubbing, Smooth Trimming), Artificial Marble, Gypsum Board, Rock Wool Sound Absorbing Tex, Vinyl Tile, SGP, Water-Paint on Gypsum Board, T15 Rock Wool Sound Absorbing Tex Design Period 2000.8–2001.9 Construction Period 2001.12–2005.8 Architecture Design Kim Sang-sik, Kim Yong-mi | G.S Architects & Associates + Kim Hyun-chul | Seoul National University Design Team Kim Se-won, Lee Seok-beom, Yu Beom-seop, Kim Je-hyeong, Kwon Min-hyeok, Jeong Cheol-u, Lee Jong-hyeok, Kim Yong-in, Moon Se-ho, Jeong Si-nae, Jeong Je-song, Jeong Hyeon-seok, Lee Jik-hyeon, Lee Jong-gyu, Kang Jun-gu, Kim Ye-hwa, Lee Seung-jong Photographer Lee Ki-hwan

South view of Jeonnam Provincial Government Office

View of the administration building, a landmark tower of the complex.

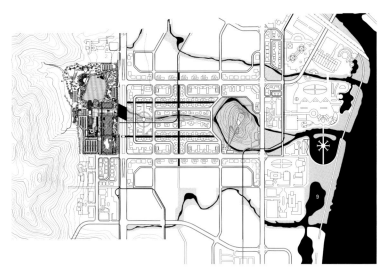

1. JEOLLANAM-DO PROVINCIAL GOVERNMENT
2. PUBLIC BUSINESS FACILITIES
3. NAMAK LAKE
4. APARTMENT & STORES FACILITIES
5. CULTURAL FACILITIES
6. CENTRAL PARK
7. COMMERCIAL FACILITIES
8. MUSEUM
9. ECOLOGICAL LAKE
10. CONVENTION CENTER

NAMAK MASTER PLAN

With preservation of Namak lake, a little stream, a field of reeds, and a pine forest, ginkgo trees, the tree of the province, were planted along the north-south green axis of Namak new town, extending its axis into the complex. While opening the green axis toward Mt. Oryong, the east-west axis of the site is open toward Namak lake. At the intersection of two axises, citizen's plaza is provided for the public as the main space on the green axis.

The roof is symbolically designed like a wing in a sharp plate to embrace the sky. The administration building erects vertically, and the civil service building is built in horizontally long mass, while the assembly building floats above the ground. Its sculptural form consists of a floating board and masses in asymmetry and dynamic proportions. In civil service building, the main conference room is located above the civil hall, while a multi-purpose lecture hall is placed in front, and the administration building is arranged right beside the civil service building. The assembly building consists of workspaces and transparent grass skins, with the citizen's plaza on the right. Two levels of administration are vertically connected for amicable atmosphere of workspaces.

The governor's office is located on the ninth floor, in the middle of all the administration offices. Offices in close relationships such as a vice governor's office and a project management offices are located on 10th~12nd floors, while there are halls on the fourth and fifth floors, leading to the civil service building. The assembly building consists of conference-assembly hall used for sessional period and office-assembly chamber for general workers and visitors. Between the two areas, there are offices for permanent committee and members of congress, and corridors. Conference-assembly hall area represents transparent image of congress democracy with transparent curved elevation, while office-assembly chamber area is designed in a rectangular shape to increase spatial efficiency and flexibility.

The citizen's plaza is open to Namak new town and Mt. Oryong, separating the administration mass and civil service mass.

View from the main approach. The civil hall is designed in a horizontal mass with horizontal louvers to alleviate its authoritative image.

1 T3 ALUMINUM SHEET / FLUOROCARBON RESIN PAINT
2 T24 PAIR GLASS
3 T50 EXTRUDED CEMENT PANEL
4 Ø30 STEEL PIPE / FLUOROCARBON RESIN PAINT
5 URETHANE PAINT AFTER CONCRETE SURFACE TREATMENT
6 FLUOROCARBON RESIN PAINT ON STEEL PIPE
7 ALUMINUM GRILL
8 T30 GRANITE SMOOTH TRIMMING

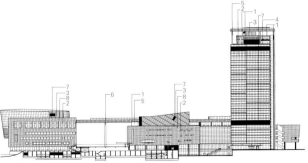

FRONT ELEVATION

RIGHT ELEVATION

REAR ELEVATION

LEFT ELEVATION

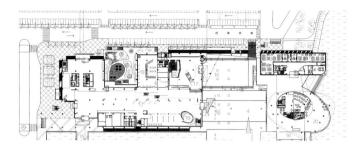

1ST FLOOR PLAN

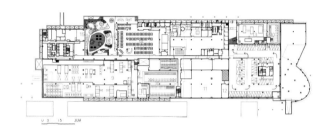

B1 FLOOR PLAN

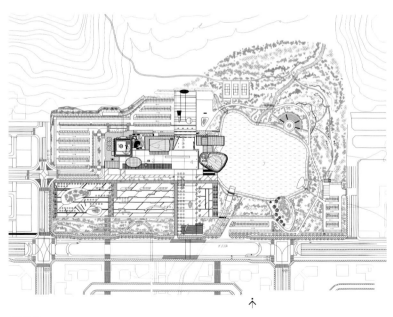

SITE PLAN

5TH FLOOR PLAN

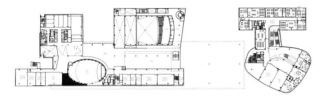

4TH FLOOR PLAN

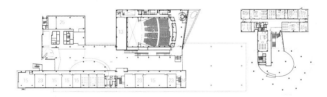

3RD FLOOR PLAN

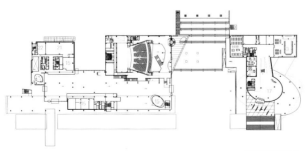

2ND FLOOR PLAN

1 ADMINISTRATION BUILDING
2 AUDITORIUM
3 CONFERENCE ROOM
4 CITIZEN'S PLAZA
5 ASSEMBLY & BUSINESS BUILDING
6 ASSEMBLY & CONFERENCE BUILDING
7 NAMAK LAKE
8 NATURAL STUDY AREA
9 MACHINE ROOM
10 ELECTRIC ROOM
11 STORAGE
12 FIRE & GAS ROOM
13 HVAC ROOM
14 CAFETERIA
15 OFFICE
16 ENTRANCE
17 HALL
18 PR ROOM
19 CIVIL SERVICE ROOM
20 TERRACE
21 CENTRAL CONTROL ROOM
22 EMERGENCY CENTER
23 ARCHIVED LIBRARY
24 RESTING AREA
25 UPPER 2F
26 PRESS ROOM
27 MAIN CONGRESS HALL
28 BROADCASTING AREA
29 INTERVIEW ROOM
30 STUDIO
31 MULTIPURPOSE ROOM

Composition of void space for civil hall and solid space for the main conference room

Overall view of civil service hall viewed from the 4th floor

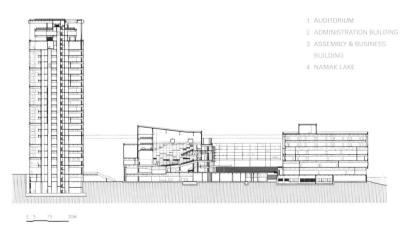

1 AUDITORIUM
2 ADMINISTRATION BUILDING
3 ASSEMBLY & BUSINESS
  BUILDING
4 NAMAK LAKE

0 5 15 30M

CROSS SECTION

View of the auditorium approaching from the civil service building. Linear stairway leads people from civil hall to the auditorium.

Void space between the auditorium and an archived library, which increases juxtaposition and spatial depth.

Connection from the chairman room of the assembly building to the main congress hall. Exterior walls and curtain walls are creating a gallery.

Connection from the office building to the main assembly hall

View of the main congress hall seen from the 1st floor hall of the assembly building.

>> OFFCIE

SEOUL CENTRAL POST OFFICE(SCPO)

Location Chungmuro 1(il)-ga, Jung-gu, Seoul, Korea Site Attribute General Commerce, Fire-Prevent, Center Aesthete, Cultural Assets Protection Site Area 6,134.80m^2 Total Floor Area 72,718.50m^2 Landscape Area 1,068.90m^2 (17.42%) Building Coverage Ratio 51.44% Floor Area Ratio 710.13% Building Scope B7~21F Maximum Height 89.82m Design Period 2003.1~2003.12 Construction Period 2003.9~2007.9 Architecture Design Lee Sang-leem, Kang Sung-in | SPACE GROUP + Jeong Young-kyoon, Jo Nam-seung | HEERIM Architects & Planners + Kim Heung-su, Kim Sang-rak | HANGIL Architects & Engineers + James R. DeStefano, Scott Sarver | DeStefano Keating Partners, Ltd. Design Team SPACE_Kim Beom-jun, Park Yun-seok, Kim Hyeon-u, Park Sang-hyeon, Lee Pyeong-ju, Lee Jong-yeong, Hwang Yun-taek, Kim Yong-tae, Lee Min-sun, Yang Seung-hyeon, Song Yeong-seok, Kim Jae-yun, Ryu Jeong-min / HEERIM_Go Won-jun, Park Jin-geun, Choi Min, Jo Il, Park Myeong-wan / HANGIL_Jeon Seong-jin Interior Design HEEHOON D&G Construction GS E&C + DAELIM Industrial + HYUNDAI Development Company + HANWHA E&C Client Ministry of Information and Communication Photographer Lee Ki-hwan

The site of Seoul Central Post Office forms a triangle with Seoul City Hall and South Gate(Soongryemun), taking an important role as a symbolic gate of the city. The district, located in the Myung-dong shopping arcade which is connected to Sogong-dong, is expected to absorb a large number of people. Thus, the public, social, and historical quality of the site, based on the temporal and extensive context had to be consistently preserved from the design phase to the landscape design phase of the project.

The entire mass is stepped back to form a connecting square with the empty sites around the rotary, so the square takes a central role as a connecting bridge or as a node for the site. The post booths for users were arranged on the first floor and the second floor, which are connected to the front square through Sunken Garden, and the external spaces such as Central Fountain Plaza at the rotary were designed to interact with the internal spaces of the building as an open garden for citizen. In addition, an urban garden and a reception center were constructed on the 10th floor to provide the users with a space for relaxation and meeting. Considering the symbolic role of the post office as a center of the city, the vertical sides of the building were designed to embrace the two towers standing in symmetry, therefore it looks as if the towers are veiled by the overlaid skin. The skin of the building, which is emphasized by the space due to the mixture of the contrary materials such as high-tech materials and the natural materials including stone and punch window, makes a great harmony with the surroundings.

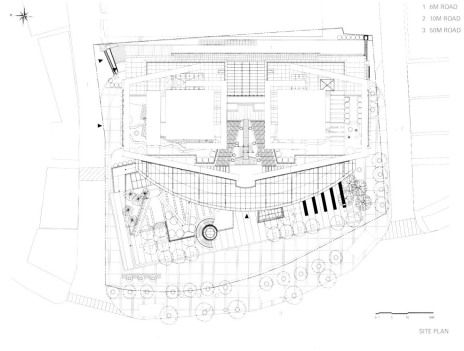

1 6M ROAD
2 10M ROAD
3 50M ROAD

SITE PLAN

SITE TRIANGLE

- Phase 1 : Land Mark
- Phase 2 : 열림, 조화
- Phase 3 : 변형
- Phase 4 : Green Network

DESIGN PROCESS

두개의 매스
- 남대문과 평행을 맞는 축 연장
- 적정 높이를 통한 쾌적한 업무환경 조성

대칭형 매스
- 강한 상징성, 중심성 표출
- 중앙의 집구성 강화

스킨(Skin)
- 사각 건물군 속의 지선형 매스로 시선유도 및 안정감 강화
- 수직적 상승감 표현

- 도심공간의 매개길 형성

MODELING

The relaxing area for citizens.

OFFICE
SEOUL CENTRAL POST OFFICE(SCPO)

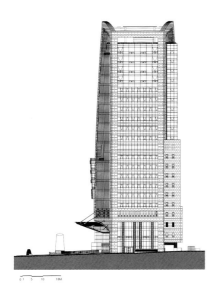
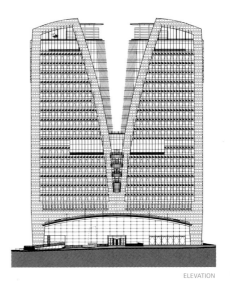

ELEVATION

1 HALL
2 OFFICE
3 POST WORKROOM
4 LOUNGE
5 GALLERY
6 LOBBY
7 POST SERVICE AREA
8 CONVENIENT FACILITIES
9 PARKING LOT
10 FITNESS CENTER
11 PR ZONE
12 CONFERENCE ROOM
13 RECEPTION CENTER

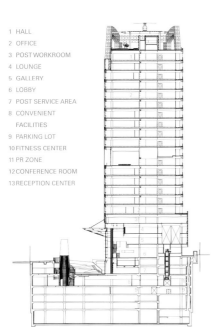
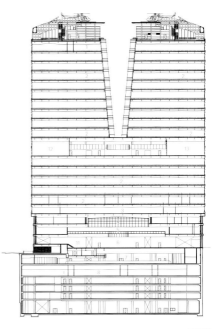

SECTION

The main hall of the 1st floor post office viewed from the 2nd floor offers a sense of openness.

Overall view of the main entrance

View of 2nd floor open space on the 1st floor

Basement entrance core of the sunken

Exhibition hall of the basement

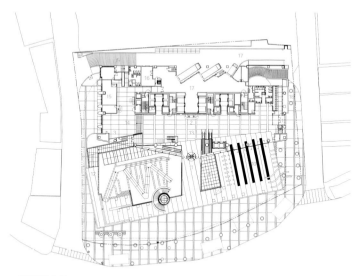

1ST FLOOR PLAN

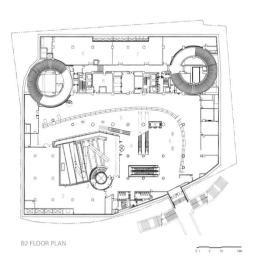

B2 FLOOR PLAN

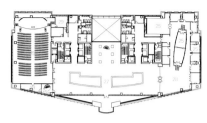

10TH FLOOR PLAN

1 SUNKEN GARDEN	15 RESTING AREA
2 INFORMATION	16 EMS RECEPTION ROOM
3 PR ZONE	17 ARRIVE & DEPART AREA
4 MEETING AREA	18 AUTOMATION WINDOW
5 POST & FINANCE SERVICE AREA	19 GALLERY
	20 MDF ROOM
6 EMERGENCY CENTER	21 HVAC ROOM
7 WORKROOM	22 POB RECEIVING ROOM
8 MARKETING ROOM	23 STORAGE
9 POND	24 CIVIL SERVICE ROOM
10 GARDEN	25 CONFERENCE ROOM
11 MAIN ENTRANCE	26 OFFICE
12 SUB ENTRANCE	27 INDOOR GARDEN
13 LOBBY	28 RECEPTION CENTER
14 STAFF'S ENTRANCE	29 WAITING ROOM

6–9TH FLOOR PLAN

2ND FLOOR PLAN

12–21TH FLOOR PLAN

12th floor, Rest area of the roof garden

12th floor, Green area of the roof garden

12th floor, Indoor of the roof garden

>> OFFICE

MERITZ FIRE & MARINE INSURANCE BUILDING

Location Yeoksam-dong, Gangnam-gu, Seoul, Korea Site Attribute General Commerce, Central Aesthete, District Unit Plan, Parking Lot Limit Site Area 4,464.1m² Building Area 1,551.36m² Total Floor Area 57,435.79m² Landscape Area 810.7m² Building Coverage Ratio 34.75% Floor Area Ratio 785.01% Building Scope B6~30F Structure Steel Exterior Finish Granite Rubbing, Aluminum Sheet, T24 Color Low-E Pair Glass Interior Finish Floor_Marble(Lobby, Typical Floor Hall), Carpet Tile(Office) / Wall_Marble(Lobby, Typical Floor Hall), Water Painting(Office) / Ceiling_Barrisol(Lobby), V.P on Gypsum Board(Typical Floor Hall), T12 Rock Wool Sound-Absorbing Tex Design Period 2001.8~2002.12 Construction Period 2002.11~2005.10 Architecture Design SHINHAN Architects & Engineers + KEATING/KHANG LLP Design Team Lee Yeong-gu, Hong Bong-jin, Gwak Mun-jin, Park Hui-jun, Park Mu-hyeon Construction HANJIN Heavy I&C Client Meritz Fire & Marine Insurance Photographer Lee Ki-hwan

Located on the intersection between Teheran road and gangnam road, the new company building for Meritz Fire & Marine Insurance is designed to become a new urban landscape, creating an impressive skyline both from the other side of the Han river and from the main streets of gangnam district. The building is intended to be a completely new place not only with functional and pleasant working spaces but also with various qualities of outdoor spaces. Its strong geometric form and 'sky window', a void below the sloped rooftop, invites vital and lively atmosphere onto the impassive and static commercial environment through interaction between sky and light. The outer skin of the building is composed of two kinds of walls; glass curtain walls for working spaces on each floor and black granite stone walls with small openings for public spaces such as resting areas, mechanical rooms, and elevator halls, etc. While the sky window is designed to be a rooftop garden, a gigantic window is repeated in the facade of the lobby on the first floor. Sceneries of the lobby area and a waterscape of cascade on the other side of the floor is vividly overlapped through transparent curved windows behind the facade. In addition to these overlapped scenes, visitors' movements in the lobby are naturally connected with the waterscape(cascade). By utilizing the land's slope, the waterscape provides an affluent resting space with water, trees, and grasses and with sense of time and quality of place to the people in the downtown. Under the sloped cascade, rental spaces are facilitated with various spatial qualities. Visitors' movements and visions from the foyer of the grand ballroom are open to the north sunken garden surrounded by a circular parking ramp, physically connected with the cascade by transparent stairs and a bridge in the sunken garden.

Two vertical masses are separated on the roof garden level.

Two vertical masses supporting the slope roof create dynamic skyline in a distant view.

Roof garden seen from the northwest

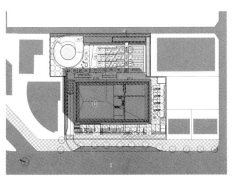
SITE PLAN

1 MAIN ENTRANCE
2 OPEN AREA
3 WATER SPACE
4 WALKER'S ENTRANCE
5 ROAD

150　　Front glass window seen from the lobby entrance on the first floor. There is a picturesque waterscape beyond the window.

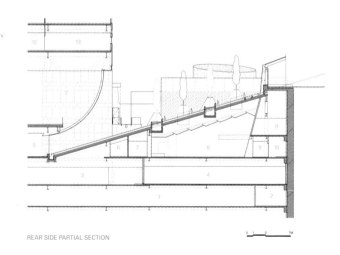

1　PARKING LOT
2　WASTE DISPOSAL FACILITY
3　PARKING PASSAGE
4　CAFETERIA FOR STAFF
5　HALL
6　LABORER'S OFFICE
7　LOBBY
8　RESTAURANT
9　CORRIDOR
10　HVAC ROOM
11　SUPPORT FACILITY
12　OFFICE
13　RESTING AREA

REAR SIDE PARTIAL SECTION

Lobby seen from the rear side. Bridge above the waterscape connects both waterscape and outdoor space in the north and the south.

Rear view from the northwest, reminds us of an urban oasis in harmony with waterscape, trees, and grasses on a sloped land.

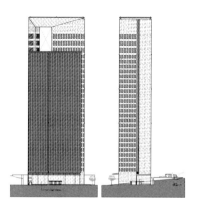

ELEVATION

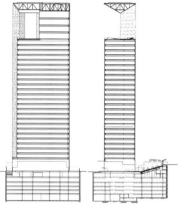

SECTION

1ST FLOOR PLAN

B1 FLOOR PLAN

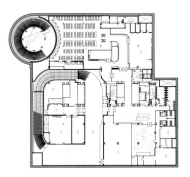

B2 FLOOR PLAN

27~29TH FLOOR PLAN

16~26TH FLOOR PLAN

3~15TH FLOOR PLAN

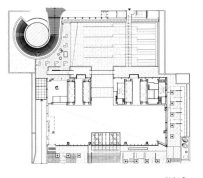

2ND FLOOR PLAN

1 CAFETERIA FOR STAFF
2 KITCHEN
3 ELEVATOR HALL
4 OFFICE
5 CONVENIENT FACILITY
6 CONTROL OFFICE
7 HVAC ROOM
8 PARKING PASSAGE
9 RESTAURANT
10 FLOWER GARDEN
11 SLOPING CASCADE
12 EMERGENCY CENTER
13 BANQUET HALL
14 MAIN ENTRANCE
15 LOBBY
16 CONFERENCE ROOM
17 ROOF GARDEN

>> OFFICE

YEOUIDO BUILDING TAEYOUNG

Location Yeouido-dong, Yeongdeungpo-gu, Seoul, Korea Site Attribute General Commerce, Central Aesthete, Height Limit, Public Facilities Protection Site Area 3,805m² Building Area 2,273.27m² Total Floor Area 41,858.8m² Landscape Area 589.69m² Building Coverage Ratio 59.74% Floor Area Ratio 712.55% Building Scope B5–13F Structure RC, SRC Exterior Finish T24 Color Pair Glass Curtain Wall, T10 Aluminum Honeycomb Panel Interior Finish Steel Wall Cladding Design Period 2003.1.1–2004.6.4 Construction Period 2004.8.20–2007.3.26 Architecture Design Byun Yong | WONDOSHI Architects Group, Ltd. Design Team Lee Gi-jeong, Kim Jin-mo, Jo Hyeon-hui, Kang Ju-seong, Lee Gyu-ho, Kim In-seong, Lee Je-seung, Kim Hyeon-a, Park Jin, Kim Hak-gi, Jeong Seung-hyeon Construction TAEYOUNG E&C Photographer Lee Ki-hwan(Except Otherwise Indicated), Park Young-chae

"A dish is made by shaping dough and the dish is useful because it is empty."
It contains routine life, sunbeam, the wind, and the rain and stars.
And. . . . it contains the universe.

The site for this project is located around the northwest entrance of Yeouido park. A series of buildings which represent their identities are arranged around the site where construction of high-rise building is restricted. The key in this project was to solve the problem of how the 228,000m² area park should interact and make a harmony with the surrounding buildings. The 2,273.27m² mass was designed in '☐' shape through many studies. By arranging the convergent core in the space blocked by adjoining buildings, it can be transformed into a vital space with support of the observatory elevator and transparent stairways. The main entrance on the first floor is smoothly connected to the underground 300-seat concert hall through piloti, thus it provides visitors with convenience and functions. Four out of all the ground floors are decorated as office while nine floors below the office area will be rental working spaces. All the exterior and internal spaces were designed to be consistent with the park, using landscape elements such as trees and plants. The depth of the standard office is 12m, and both sides of the office are covered by curtain walls for better ventilation and lighting. The vertical sides of the building were emphasized by its unique lines and angles. It was built using Top-Down technology and slurry wall for soundproofing and reduction of air flow. And sunlight absorbtion controllers were properly installed to reduce thermal Load of HVAC system due to increase of the surface area. We dream of 'hope' in the space where is brilliant, clear, and vital.

The building is accessible through the piloti.

The facade viewed from the road

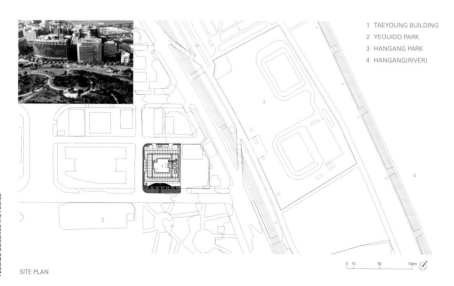

1 TAEYOUNG BUILDING
2 YEOUIDO PARK
3 HANGANG PARK
4 HANGANG(RIVER)

SITE PLAN

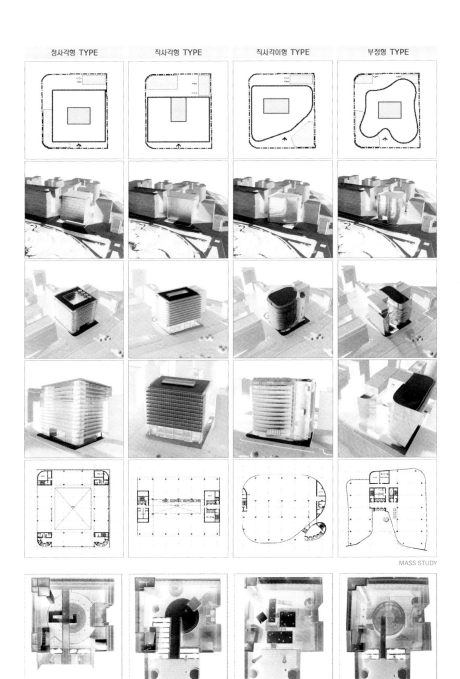

MASS STUDY

SUNKEN GARDEN STUDY

The lobby on the 1st floor

1 OFFICE
2 PARKING LOT
3 COURTYARD
4 MAIN ENTRANCE PILOTIS
5 SUB ENTRANCE PILOTIS
6 CONCERT HALL
7 STAFF'S CAFETERIA
8 CAFE
9 LOBBY
10 BANK
11 HALL
12 CAFETERIA
13 MACHINE ROOM

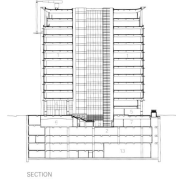
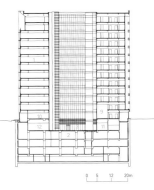

SECTION

General view of the main entrance

The landscape element has a visual connectivity with Yeouido Park in the vicinity.

The courtyard is a place where the bright and vigorous design of the building can be experienced at the maximum level.

The upper zone viewed from the courtyard of atmospheric open type combined with open and closed

The visitors are naturally guided to the underground courtyard.

The cafeteria of the courtyard

The corridor of the office

OFFICE
YEOUIDO BUILDING TAEYOUNG

The voided Staircase is connected to the outside.

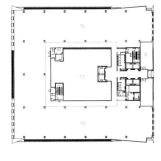

1 HALL
2 CONCERT HALL
3 PRACTICE ROOM
4 GENERAL SHOP
5 CAFETERIA
6 HVAC ROOM
7 EMERGENCY CENTER
8 COURTYARD
9 CAFE
10 MAIN ENTRANCE PILOTIS
11 LOBBY
12 BANK
13 OPEN
14 SUB ENTRANCE PILOTIS
15 OFFICE
16 UPPER LOBBY

1ST FLOOR PLAN

3RD–11TH FLOOR PLAN

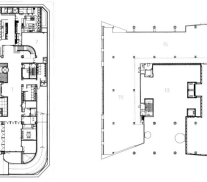

B1 FLOOR PLAN

2ND FLOOR PLAN

>> OFFICE

EMBASSY OF THE REPUBLIC OF KOREA IN CHINA

Location Beijing, China Site Area 15,940.00m² Building Area 4,139.83m² Total Floor Area 16,294.60m² Landscape Area 6,002.00m² Building Coverage Ratio 25.97% Floor Area Ratio 63.10% Building Scope B1-5F Structure RC Exterior Finish Granite, T18 Color Pair Glass, Aluminum Sheet Design Period 2003.10-2005.12 Construction Period 2003.9-2006.9 Architecture Design Lee Dong-hee | KUNWON Architects Planners Engineers Design Team Jo Yeong-bum, Park Hye-ri Construction SAMSUNG C&T Photographer Kim Tae-oh

Night view of Embassy of the Republic of Korea in China from the front gate.

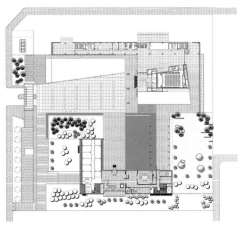
SITE PLAN

Night view of the 1st government building annex from the courtyard of the 2nd government building.

This project should reflect the Korean Culture since the Korean Embassy building was planned to be constructed in China. It means the building should contain Korean traditions on one hand and reflect Chinese traditions as well as the features of Yang magyo around the building on the other hand.

This project reflects the spatial features of the traditional Korean architecture. One of the significant features of the traditional Korean architecture is the courtyard. The Korean courtyard is small and cozy, and it connects a building with another building within the circle. Thus, the architect categorized this project systematically in order to construct a private court for each building.

In addition, the architect added the 'pure and brilliant images of the Far East' to the building and applied 'Morning Calm' concept to the entrance court as a keyword. This court is in a shape of a rectangle just as the long mass in the office building that looks extremely clear and sophisticated. The floor was covered with white granite and lawn, and the wall was made of transparent double-glass to maximize the spacious effects. The most outstanding space in the building is the atrium that serves as a lobby. This is a public space in the office building, which will be built in the concept of the main floored room in a traditional Korean house. As a living room, the floored room is half-inside and half-outside space which connects the front court to the backyard. It is an open space located between the office building and the affiliated building in which both buildings can be seen from the main entrance.

Although the original courtyard plan was withdrawn due to the extension work for the second office building in accordance with the change of use for the building, the architect constructed two courtyards to complement a function of the office buildings and to create a distinguished outdoor space.

While the entrance courtyard can be compared to 'Morning Calm', the court before the second office building reminds us of an inner court in a traditional Korean house. The architect built a pond in the court and built a reception/meeting room in the pond as a pavilion. He emphasized the function of the court as a bridge by putting the court in the center of the second office buildings in both visual and functional way.

The double skinned facade of the 1st government building.

The atrium of the 1st government building corresponding to the main floored room that separates the working area from the supporting area.

>> OFFICE
ENVISION

Location Gasan-dong, Geumcheon-gu, Seoul, Korea Built Area 296m² Interior Finish Floor_Polished Tile, Mosaic Tile, Wood Flooring, Carpet Tile / Wall_Color Glass, Mosaic Tile, Magnetic Paint / Ceiling_Barrisol, Paint Completion 2007.12 Interior Design Lee Mi-young, Gwon Hyuk-geon | Design Pure Design Team Moon Kyu-do, Jung Yun-mi, Yun Soo-yeon Client Envision Co., Ltd. Photographer Kim Jae-yun

1	ENTRANCE
2	STORAGE
3	APP. MOSAIC TILE FIN.
4	T5 MIRROR GLASS FIN.
5	APP. COLOR LACQ. FIN.
6	T5 BACK PAINT GLASS FIN.

General view of the entrance

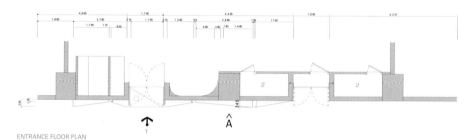

ENTRANCE FLOOR PLAN

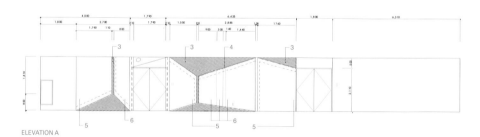

ELEVATION A

They have an office that is currently being used. But for a separate purpose, the office was extended. While still conducting pre-existing office work, the extended office was hoped to be a space where employees could relax and exhibit creativity enjoying new spaces such as convenient facilities, lounge, hobby and education places. Also they wanted a freer, unrestricted communication atmosphere in the company. Considering this, partitions were minimized in this open, sharing space "Need" By opening each space that could be specifically used, an atmosphere like an open cafe has been created. The prejudice that meetings should be held in a meeting room was thrown away.

Humorous space names such as Idea incubator, Idea bay, Study room, Book cafe, and Igloo may seem confusing when the purpose of the space is relaxing, meeting, studying or reading, but in fact, those names are the expression of the company's thoughts considering "Idea" the most important. This place is not a company in which all the energy is used up by working, but a space for working, refreshing and growing. Instead of staying in a line at Starbucks every morning, employees would have simple food and drink coffee in the company lounge listening to music. And then, they would remind the main schedule of company through PDP installed in the ceiling, and start to share opinions with each other in a comfortable posture, not sitting on hard office chairs. In this open space, it is possible for the Technology team passing by to join the Marketing team in a meeting.

1 SIZE : 2,410X1,885
2 APP. VENNER ON T1.2 PLATE FIN.
3 MAGNET PAINT(BLACKBOARD PAINT MIX) ON T1.6 PLATE FIN.
4 ☐-45X76 ST'L PIPE
5 APP. VENNER ON T1.2 STEEL PLATE FIN.
6 T4 PLATE STEEL(W=79)
7 MAGNET PAINT(BLACKBOARD PAINT MIX) ON T3 PLYWOOD 4PLY FIN.
8 APP. VENNER ON T3 PLYWOOD 4PLY FIN.
9 T18 PLYWOOD
10 ☐-30X30 DOUGLAS FIR SQUARED TIMBER
11 MAGNET PAINT ON T3 PLYWOOD 4PLY FIN.
12 T4 PLATE(W=76)
13 T9 MDF(W=76)
14 T12 MDF(W=100)
15 T9 PLYWOOD
16 APP. VENNER ON T12 PLYWOOD FIN.
17 PAINTING ON T12 PLYWOOD FIN.
18 DECO LAMP LIGHTING COLOR
19 WELDING AFTER BOLT FIXING USE ┐ SHAPE STEEL CLAMP
20 SIZE : 2,700X1,800
21 REINFORCEMENT PIPE(OUTER) SIZE : 1,500X800
22 OPEN

Idea incubator B type

The round steps of the 3m high stairway serve as seats, podium for the speaker

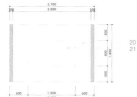

IDEA INCUBATOR A TYPE TOP VIEW

IDEA INCUBATOR B TYPE TOP VIEW

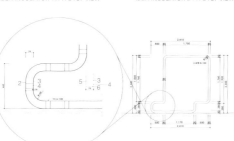

IDEA INCUBATOR B TYPE FRONT VIEW

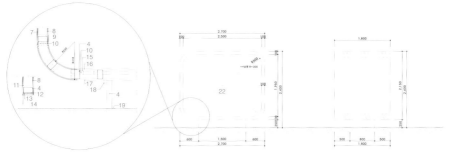

IDEA INCUBATOR A TYPE FRONT & SIDE VIEW

All the spaces are open, and anyone are invited the meeting and join in

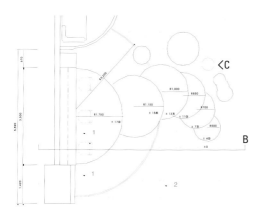

1. APP. CARPET TILE FIN (500X500)
2. APP. POLISHED TILE FIN (600X600)
3. APP. COLOR LACQ FIN (COLOR: BLACK) / APP. CARPET TILE FIN (500X500) / APP. MOSAIC TILE FIN.
4. APP. MOSAIC TILE FIN.
5. APP. MOSAIC TILE FIN. / APP. COLOR LACQ FIN (COLOR: IVORY) / APP. COLOR LACQ FIN. (COLOR: BLACK)
6. TILE CORNER MODELING(9MM ROUND)
7. T5 PLYWOOD 2PLY / APP. MOSAIC TILE FIN.
8. □-40X40 STL PIPE
9. APP. CARPET TILE FIN.(500X500) / T12 PLYWOOD / T9 PLYWOOD / □-40X40 STL PIPE
10. □-40X40 STL PIPE / T9 PLYWOOD / APP. COLOR LACQ FIN.(COLOR: IVORY)
11. WELDING AFTER BOLT FIXING USE ㄱ SHAPE STEEL CLAMP
12. APP. COLOR LACQ FIN.(COLOR: BLACK) / T22 MDF / APP. COLOR LACQ FIN.(COLOR: YELLOW)
13. APP. COLOR LACQ FIN. (COLOR: BLACK) / ARTIFICIAL
14. APP. COLOR LACQ FIN (COLOR: IVORY)
15. APP. MOSAIC TILE FIN. / APP. COLOR LACQ FIN (COLOR: BLACK)
16. APP. COLOR LACQ FIN (COLOR: ORANGE)

BOOK CAFE FURNITURE TOP VIEW

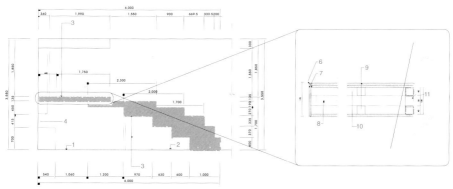

FURNITURE SECTION B

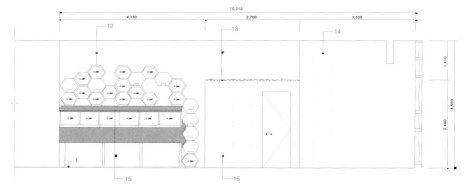

BOOK CAFE ELEVATION C

The conference room viewed from the book cafe

1 OPEN
2 Ø85 PAPER PIPE
3 APP. VINYL PAINT FIN.
4 T3 APP. MOSAIC TILE(H=260MM) / T5 MIRROR(H=45MM) / VENEER WOOD ON T5 PLYWOOD 2PLY(H=105MM) FIN. / T8 COLOR GLASS(H=1,310MM) / VENEER WOOD ON T5 PLYWOOD 2PLY(H=250) FIN. / T3 APP. MOSAIC TILE(H=590)
5 □-40X40 ST'L PIPE
6 EXISTING CONCRETE SLAB, APP. VINYL PAINT FIN.
7 T8 TEMPERED GLASS
8 T13 APP. POLISHING TILE FIN.
9 T5 PLYWOOD(W=300MM)
10 CEMENT MORTAR
11 EXISTING CONCRETE SLAB
12 ALUMINIUM MOULDING(H=5MM)
13 □-30X30 MISUNG WOOD(@900)
14 T9.5 GYPSUM BOARD 2PLY, APP. VINYL PAINT FIN.
15 T12 PLYWOOD(W=88MM)
16 APP. MOSAIC TILE(H=260) FIN.
17 T5 MIRROR(H=45MM)
18 VENEER WOOD ON T5 PLYWOOD 2PLY(H=100MM) FIN.
19 T8 COLOR GLASS(H=1,300)
20 VENEER WOOD ON T5 PLYWOOD 2PLY(H=250) FIN.
21 T3 APP. MOSAIC TILE(H=590)
22 T9.5 G.B 1PLY
23 T9 MDF
24 □-50X50 S'STL PIPE(@450)
25 □-50X50 S'STL PIPE

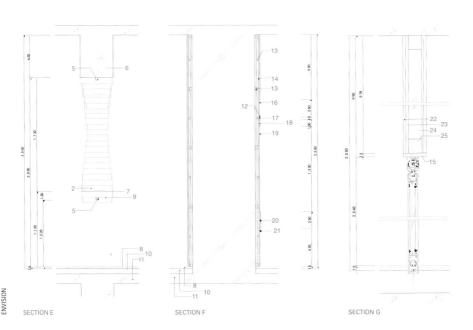

SECTION E   SECTION F   SECTION G

Conference room_partition wall close　　　　　　　　Conference room_partition wall open

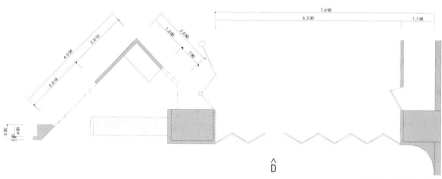

CONFERENCE ROOM PARTIAL FLOOR PLAN

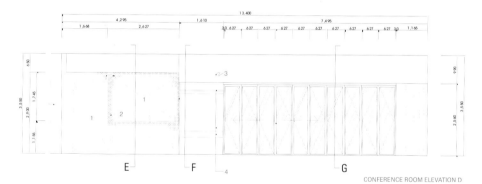

CONFERENCE ROOM ELEVATION D

1 T12 TEMPERED GLASS FIN. / T5 PLYWOOD 1PLY, APP.
  COLOR LACQ FIN.(COLOR= IVORY)
2 T9.5 G.B 1PLY, APP. COLOR LACQ FIN.(COLOR=BLACK)
3 EMBO SHEET / T12 TEMPERED GLASS FIN.
4 T9.5 GB. 1PLY
5 T9 MDF
6 □-50X50 S'STL PIPE(@450)
7 □-50X50 S'STL PIPE
8 T12 PLYWOOD(W=68MM)
9 PAINT ON T9.5 GB. 1PLY FIN. (COLOR=BLACK)
10 □-30X30 SQUARED LUMBER WOOD / T5 PLYWOOD /
   T5 PLY-WOOD 2PLY(W=80MM)
11 T5 PLYWOOD
12 □-30X30 SQUARED LUMBER WOOD
13 T5 PLYWOOD
14 T9 MDF(W=35)
15 T5 PLYWOOD(H=26MM)
16 T9 MDF(W=20MM)
17 FL32W FLUORESCENT LAMP COLOR(T=120 @250)
18 BARRISOL
19 PAINT ON T9.5 G.B 1PLY FIN. (COLOR=BLACK)
20 T18 PLYWOOD(W=146MM)
21 T18 PLYWOOD(H=85MM)
22 □-30X30 SQUARED LUMBER WOOD
23 EMBEDDING TYPE ELECTROMOTION SCREEN (100")
24 T12 PLYWOOD

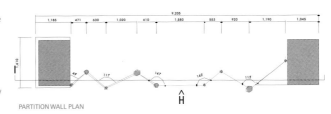

PARTITION WALL PLAN

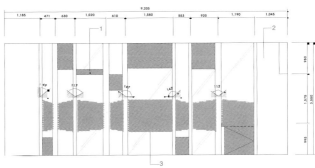

PARTITION WALL ELEVATION H

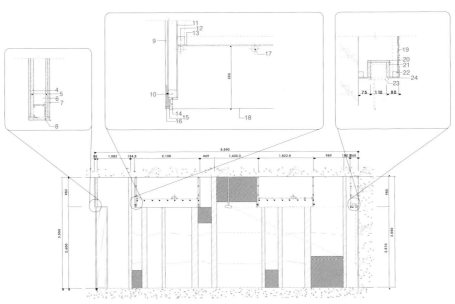

PARTITION WALL SECTION I

1 ENTRANCE
2 CAFETERIA
3 CONFERENCE ROOM
4 WORKING SPACE
5 BOOK CAFE
6 BILLIARD ROOM
7 SLEEPING ROOM

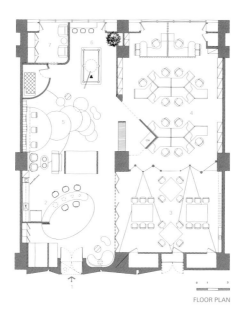

FLOOR PLAN

Lobby

\>\> OFFICE

BLOOMING LOUNGE

Location Hannam-dong, Yongsan-gu, Seoul, Korea Built Area 745m² Interior Finish Floor_Corian, Wood Flooring / Wall_F.R.P, Extenzo Design Period 2006.1~2006.3 Construction Period 2006.4~2006.5 Interior Design Kang Shin-jae, Choi Hee-young | VOID planning Design Team Oh Hwan-woo, Byun Jin-sung, Chung Jang-min Client Cheil Communications Photographer Kang Shin-jae

The shapes of trees in full bloom reaching up to the sky

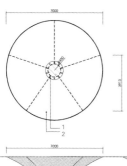

1 APP LIGHTING INSERT ON ST'L PIPE FRAME
2 EXTENZO LIGHTING SHEET FIN. ON ST'L PIPE FRAME
3 WHITE PAINTING FIN.

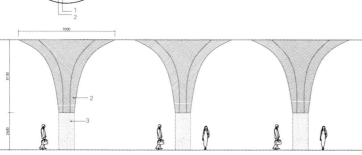

OFFICE
BLOOMING LOUNGE

POST ELEVATION

The deck of the open-space terrace is extended to the lounge, making the visitors in the lobby feel as if they are in an outdoor space.

The lobby space of the headquarters building of Cheil Communications, a total marketing communications group, is designed with the concept of an 'imaginative space.' The structural frames - the mezzanine, the columns lined in the centre of the lobby, the long and narrow rectangular plan, the high floor height - are born afresh through a new concept. The structurally inevitable columns in the centre are transformed into the shapes of trees in full bloom reaching up to the sky, implying the 'blooming' of an enterprise. A rhythmic and modern atmosphere is created by finishing the walls with layers of F.R.P. The ceiling of the mezzanine is finished with mirror, adding more depth to the floor height. The lobby space is divided into the areas of reception, exhibition, lounge and an open-space terrace. The deck of the open-space terrace is extended to the lounge, making the visitors in the lobby feel as if they are in an outdoor space. Through the overall organisation, the designers intended the limited space of lobby to become a place where corporate philosophy is felt with the warmth of humanity.

A rhythmic and modern atmosphere is created by finishing the walls with layers of F.R.P.

1 LOUNGE
2 RECEPTION
3 IMAGE WALL
4 TERRACE

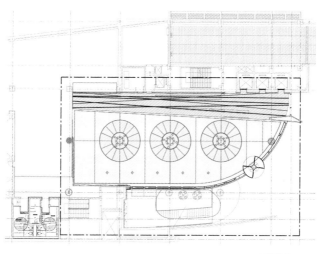

CEILING PLAN

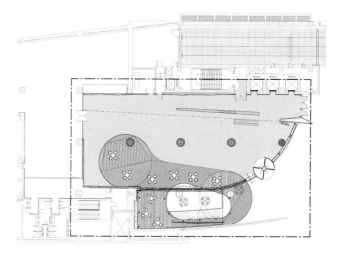

FLOOR PLAN

*OFFICE COMPETITION

\>> OFFICE

GOVERNMENT COMPLEX IN MAC

Location Jongchon-ri, Nam-myeon, Yeongi-gun, Chungcheongnam-do, Korea Site Area 122,311m² Building Area 72,487m² Total Floor Area 209,190m² Building Coverage Ratio 59.3% Floor Area Ratio 171.0% Building Scope B1~7F Structure Steel, RC Exterior Finish T24 Pair Glass, T12.3 Laminated Glass, Wood Louver, BIPV Parking 1,150 Cars Architecture Design Park Young-kern, Kim Myung-hong, Jang Deok-chan | BAUM Architects, Engineers & Consultants, Inc. + Kim Kwan-seok | ARTECH Architects + Kim Hong-ill | Dongguk University Design Team Lee Heung-sun, Kim Ho-yeong, Ju Ri-a, Im Jeong-won, Choi Jeong-bae, Jo Jeong-hwan, Park Jae-hyeon, Kim Won-yang, Kang Seong-il

LAYOUT CONCEPT: FLOATING WALL The floating wall is a concept originated from the layout of the government complex master plan, which has similarity with the image of the historical city Seoul and Suwon Hwawon Fortress. The architect analyzed the ancient map of Seoul and Suwon to reinterpret the significance and function of the fortress and introduce the result to the design. The strolling on the fortress was conceptually connected to Sunseongnori (Strolling around the fortress. Refer to Hangyeongjiryak.) In this project, the government building mass becomes a floating wall that protects the central public facilities with ready access by the users.

EXTERIOR SPATIAL PLAN: MUTILAYER DESIGN Extended from the conservation green zone on the north, the outer space proposes a green flow of the master plan. It consists of a pedestrian square linking the central administrative town, the hill areas above it and a waterscape flowing running along the ground. The roof, in particular, was designed in many layers piled upon one another, enabling the people who walk on it to provide a view on the inside and outside as well as an ecological and musical experience. The elevation plan, designed based on the motive of a transparent fortress, reflects the image of a wooden fence on the southern facade and that of a stone castle on the north facade. The variety and the sense of rhythm were realized by the appropriate separation of the mass and the variety in material and color. The southern facade with dual layer exterior system consists of two layers of different design that embody the spatiality and cubic effect of the elevation through overlapping. Metal, lighting of various patterns and LCD panels were used on the lower elevation of the piloti based on the concept of the floating wall, creating a changing bright facade.

ENERGY SAVING PLAN : SUSTAINABLE LIFE For sustainable life, the use of fossil fuel is minimized in the aspect of the environment by introducing a system utilizing the solar and geothermal power and gray water. The solar panels installed on the roof and the BIPV(Building Integrated Photo-Voltaic) film applied to the elevation gather solar energy, and the cool tube system and geothermal pipes have been planned for the active reduction of energy consumption. The eco shaft installed in the office at the centered corridor enables natural ventilation when air-conditioning is not needed, and the wildflower box that can contain freshwater can be maintained by rain, demanding the least amount of water.

SECURITY PLAN The architect thought out a perfect plan in security for emergency such as a terror attack by separating the circulation patterns for government officers, citizens and service. For the reason of safety, the parking facilities except that of the officers were located on the ground and upper levels of the building, and the entrance lamps guiding the vehicles to the parking lot in the building were centered to a single zone for efficiency in checking.

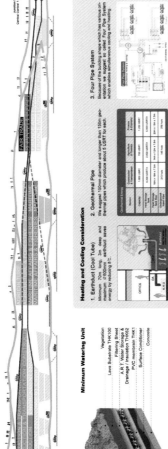
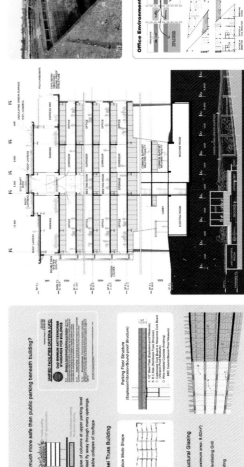
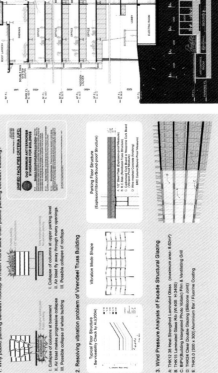
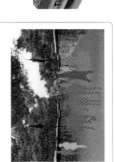

Sunseong|巡城| Picnic:

In spring and summer, people who live in Seoul used to go for a walk together on top of Seoul wall enjoying the scenery of inside and outside of the wall. It took a day for hiking the full circle trail. We call it Sunseong(Wall-Hiking) Picnic.

- Hangeongjiryak(漢京識略) by Yuboriye. 1890

Minimum Watering Unit

- Vegetation
- Lava Substrate THK100
- Filtering Sheet
- A.R.T. Water Storage & Drainage / Insulation THK62
- PVC membrane THK1
- Surface Conditioner
- Concrete

Heating and Cooling Consideration

1. Earthduct (Cool Tube)
Minimum 70m long, 3m deep and maximum ø600mm earthduct saves energy by reducing 8℃.

2. Geothermal Pipe
We suggest 150mm diameter and longer than 150m geothermal pipes which produce about 5 USRT for each.

3. Four Pipe System
Because of the building shape which has various orientation, we suggest so called 'Four Pipe System' which enables simultaneous cooling and heating.

Office Environment

Structural Consideration

1. Why public parking beneath rooftop is much more safe than public parking beneath building?
- I. Collapse of columns in basement
- II. Air blast makes progressive collapse
- III. Possible collapse of whole building

vs.
- I. Collapse of columns at upper parking level
- II. Air blast fly away through many openings.
- III. Possible collapse of rooftops

2. Resolving vibration problem of Virendeel Truss Building
- Typical Floor Structure - Serviceability Check by AIJ(2004)
- Vibration Mode Shape
- Parking Floor Structure (Explosion/Vibration/Sound-proof Structure)

3. Wind Pressure Analysis of Facade Structural Glazing
A: THK12 38 Heat Strengthened Laminated Glass (maximum area- 8.62m²)
B: THK15 Laminated Glass Rib (W:405 H:3400)
C: BIPV(Building Integrated Photo Voltaic) Film & Ventilating Grid
D: THK24 Clear Double Glazing (Silicone Joint)
E: THK3.0 (300 x 300) Aluminium Bar / Fluorine Coating

W
191

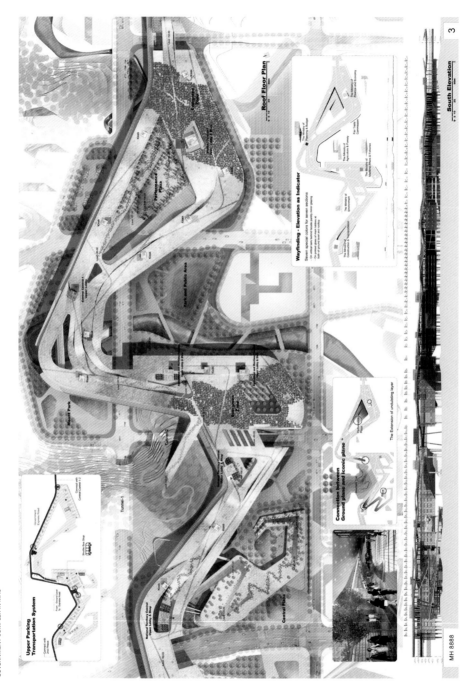

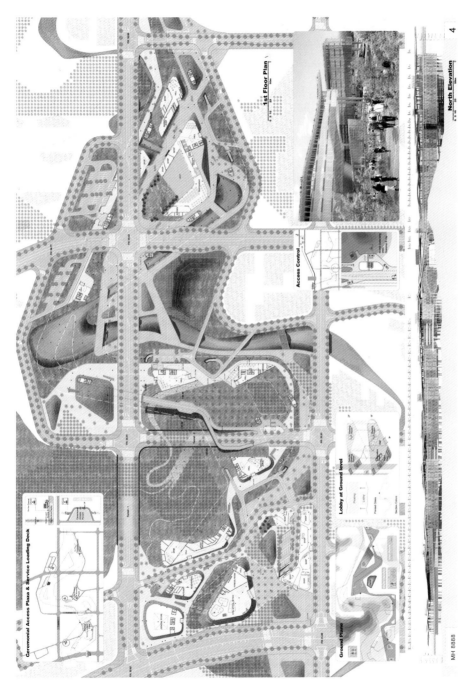

\>> OFFICE

GOVERNMENT COMPLEX IN MAC

Location Jongchon-ri, Nam-myeon, Yeongi-gun, Chungcheongnam-do, Korea Site Area 122,311m² Building Area 45,568m² Total Floor Area 214,010m² (Ground 167,370m², Underground 46,640m²) Building Coverage Ratio 37.2% Floor Area Ratio 214% Building Scope B1–8F Maximum Height 63m Structure SRC Architecture Design Jeong Young-kyoon, Kum Doo-yeon | HEERIM Architects & Planners Design Team Kim Sang, Ryu Moo-yeol, Lee In-su, Ryu Hang-soo, William Craig, Jeong Yong-joo, Kim Hae-jin, Lee Chung-mi, Yoo Sun-a, Cha Ju-yeong

LANDSCAPE We redefined the vertical zoning of the entire building in order to revitalize the roof garden. The master plan winning project attempted to maintain the intimate relationship with the roof garden by planning a visitors' parking lot on the top level. However, the parking lot was located on the ground and underground level due to the problems regarding the accessibility of vehicles, security of government buildings and the revitalization of the rooftop, leaving the top level occupied by welfare and convenient facilities. This makes the project a kind of a special mall accessed and used by not only the employees but also the visitors of the rooftop garden. The diverse level difference of the roof naturally results in dynamic and interesting spaces. On the lower part are a staircase, elevator hall and multiple convenient facilities directly linked to the top and ground level, and the upper part is an observation park. The routes connecting the roof and the top level are many in terms of variety and quantity. The direct paths for the public are the slope way that makes the best of the natural topography, moving walk core linked to the entry plaza on the west, vertical core resulting from the joints of each department, and the elevators installed in the nearby lobbies of the departments. On the ground level, 7-meter wide pedestrian axis was planned along the front of the building, reinforcing the pedestrian flow between the blocks. Running along with the pedestrian path are the lobbies of seven departments, vertical path leading to the rooftop garden and bike paths. An oblique over bridge was installed where they meet outer road so that they don't intersect with vehicle flows.

OFFICE ENVIRONMENT Each seven department has independent main lobby and entry. Functioning as an atrium, the lobby is the center of the pedestrian flow in a department. As for the design of the upper office space, the systematic spatial composition was introduced to solve the problem related to efficiency and variability. The 12-meter span workspace designed based on various technical data, 3-meter wide support zone, 2.4-meter width corridor and public conference room blocks are faced by and combined with the atrium or the courtyard, creating a single modular system. A vertical pedestrian flow and deck were installed in the vicinity of the horizontal bridge for the public. The separately planned staircase and the elevator are the public core directly leading to the roof or the top floor from the ground floor level, while the decks on each floor connected horizontally to the core function as a relaxing space as well as an overpass. The half-story high level difference of the internal linking bridge and the deck also ensures the security.

SUSTAINABILITY The effective forestation of the roof reduces the load of cooling and heating up to approx. 12% and the excessive solar irradiance in summer as well as preventing the direct damage from heavy raining. The direct sunlight flowing in the atrium can be positively used such as in energy reduction of the interior lighting and in ventilation of the inside the building through the natural air convection resulting from the temperature difference between the lower and upper spaces. The gigantic exterior courtyard planned parallel to the atrium enhances the lighting and ventilation of the upper space and the physical environment of the piloti space on the lower side.

WAYFINDING The color code classified by department is the basic way of giving the unity and independence to the overall elevation. Also, the decentralized lobby atriums ensure the sense of independency and recognition of the departments. In addition, several architectural elements were added to develop the identity and wayfinding plan of each department. Firstly, the outer skins with the same texture of the elevation were applied to the exterior wall of each lobby of the seven departments, allowing people to perceive them even from a far distance. The sign boards were installed on the exterior core wall near the lobby. Also, the color rib of the elevation was partly protruded to heighten the recognition of the lobby.

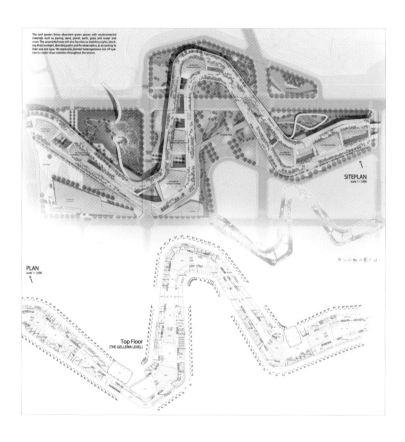

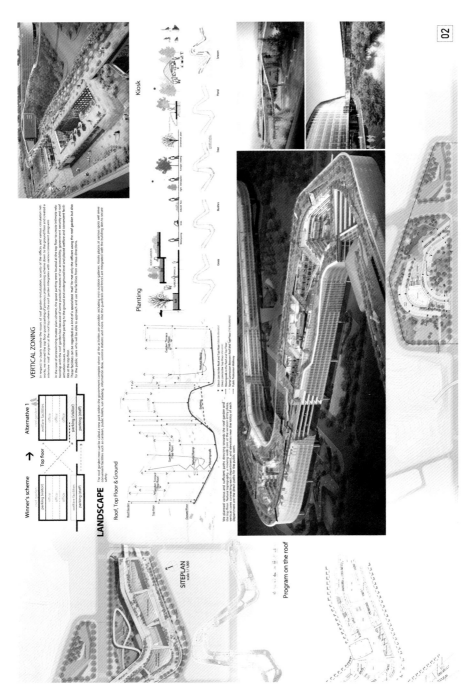

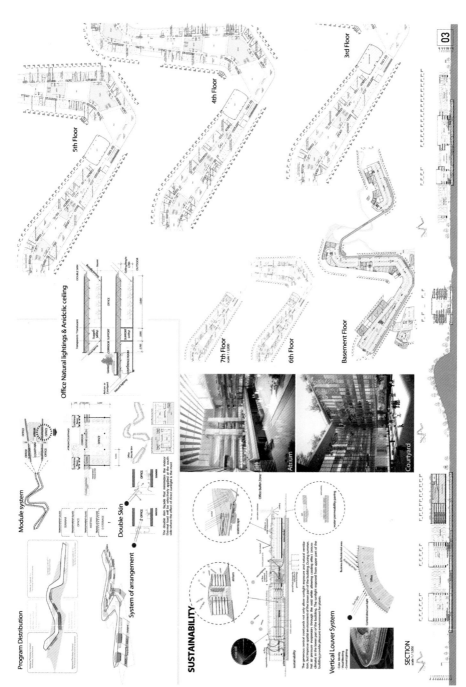

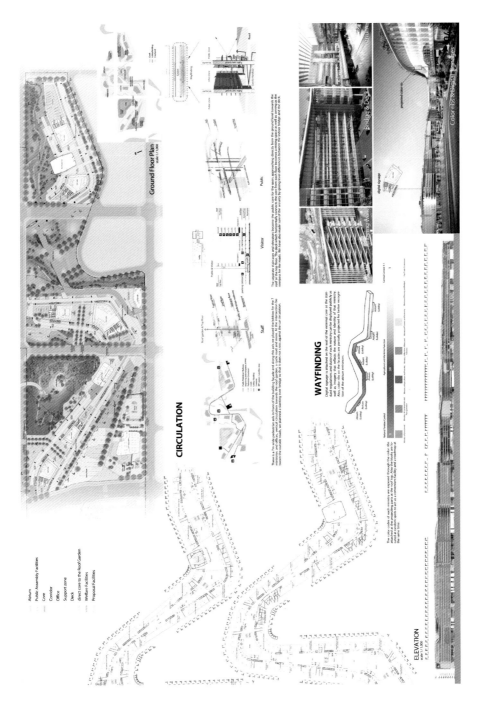

FORMing TALUS
WALKing STRIPS
FINDing CELLS

\>\> OFFICE

GOVERNMENT COMPLEX IN MAC

Location Jongchon-ri, Nam-myeon, Yeongi-gun, Chungcheongnam-do, Korea Site Area about 122,388m² Building Area 49,227m² Total Floor Area 214,740m² Building Coverage Ratio 40.2% Floor Area Ratio 175.5% Structure RC, Steel Exterior Finish Double Skin, Brick Facade Parking 1,105 Cars Architecture Design Lee Sang-leem | SPACE GROUP Design Team Cha Yong-uk, Nam Uk-jae, Kim Jin-cheol, Kim Tae-yeong, Im Ji-eun, David Fisher, Lee Yun-hui, Han Jong-beom, Go Yun-seok

The context of a virtual city is defined by Flat City, Link City and Zero City on the master plan and a continuous tangential line that constitutes the three ideas, which create a new interrelation among landscape (natural element), architecture and urban substructure on the site. The three elements - landscape, architecture and urban substructure - is the starting point of this proposal and will be the result of their interrelation.

NATURAL ELEMENT FORMing Talus (The line connecting the points of the slope and the plain)_Talus forms a base of the character of the public building and the public democratic space in designing building and natural elements. Traditionally, an oblique site on which a building, especially that of government, requires a green space or a wall for the efficient land use and the safety of the slope. We propose to break away from the existing design by positioning the building on the slope, leaving the plain as an outdoor public space thus emphasizing the significance of the public space for a new democratic government building. The architecture and nature meets at the ridge of the mountain, not going against the natural flow of the existing topography. Talus was located on a plain where the interval of the contour line is 8.1m (basic space for an office) at the minimum. Four lines can be found on the site for government buildings in the first stage of the project. The resulting location of the buildings and the plain generated in front will present the blueprint of an open democratic government building.

THE ELEMENTS AS URBAN SUBSTRUCTURE WALKing Strip_There are many elements constituting urban structure, those that guide pedestrians on the strips of the proposal play an important part in connecting various outer environments such as a park, plaza, and an area leading to the upper part of buildings as well as programs for public activities. Two of the most distinct natural elements on this site are a mountain on the west and waterscape on the east. The strips are devices to link these two significant natural elements. Each strip defines and connects nature, surrounding environments and the characters of buildings. The basic structure consists of hard and soft parts that take turns to embrace nature. In the soft part, needle-leaf trees and broad-leaved trees generate a sense of east-west axis, and at once function as a device filtering the circulation pattern in the north-south direction. While the hard parts embedded between the soft parts propose the direction of circulation toward the open space, leading people from a small private forest to a large open space. Two strips in the hard area are 'connecting strips' that link the east and west across the buildings. Each strip connects not only the landscape on earth but also the upper parts of the buildings, involving the office promenade on the upper level in an organic way, playing the role of a basic structure in creating an environmental image of the overall government building site.

ARCHITECTURAL ELEMENT FINDing Cell_The third element of the proposal begins with questions about an interface: the accessibility to the building, the conditions of the building and the outside environment (topography, trees, water and etc.) and the entry inside the building interior. The element of a cell is where the public areas of buildings and outside meet. By locating the public function on the inside, it creates spaces for citizens and gives identity to each individual government building. These cells are the founding stones that define nature, urban substructure and the buildings. The strips interact with the cells, functioning as a guide line defining the location of the cells. The 12 cells have all different characters and become the main entrance, vertical circulation paths and public space inside the buildings. The twelve exterior characters are represented as a parking area, pedestrian plaza, waters edge area and a plaza. The characters represent the identity of each department and different functions.

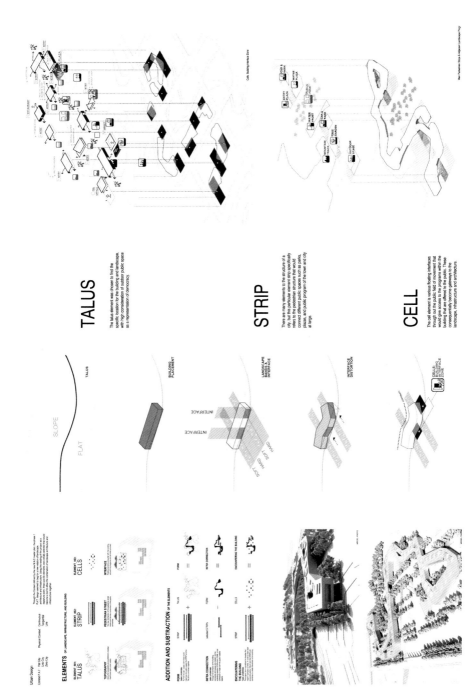

\>> OFFICE

MULTIFUNCTIONAL ADMINISTRATIVE CITY HALL

Location Hotan-ri, Geumnam-myeon, Yeongi-gun, Chungcheongnam-do, Korea Site Area 41,661.00m² Building Area 11,597.96m² Total Floor Area 33,759.42m² (Ground 26,753.72m², Underground 7,005.70m²) Building Coverage Ratio 28.85% Floor Area Ratio 66.55% Building Scope B1~6F Maximum Height 35.2m Structure SRC, RC, Steel Exterior Finish Curtain Wall, Aluminum Panel, Color Pair Glass Architecture Design Jeong Young-kyoon, Lee Byung-koo | HEERIM Architects & Planners + Kim Jung-gon | Konkuk University Design Team Lee Sang-beom, Seo Jun-bae, William Craig, Lee Jeong-won, Lee Ju-ho, Choi Ji-yeong, Hwang Jong-ho, Park Bo-yeong Client Multifunctional Administrative City Construction Agency

The competition for multifunctional administrative city hall is on the extension of environment-friendly and decentralized projects related to the multifunctional administrative city which have repercussion over the world of Korean architecture. It was a project of difficulty that has to propose a new paradigm for a city hall and reflect the growing character of the city until year 2030.

Since the site was located on the intersection of the water axis of Geumgang(River) and the green axis connecting Bihaksan(Mt.) and Wonsusan(Mt.), the relation of the two axes were interpreted into an architectural language. To meet the demand of the guideline of the design competition, which was to create an open space for citizens, the architect planned a plaza for the city in connection to the existing one on the south of the site, through which the publicity the government building was maximized. Also, a deck was installed on the lower part of the building, which can extend the flow of the plaza, so that people might feel as if they were walking on a strolling path leading to the city hall. Such a spatial flow is naturally continues to the Geumgang waters edge park and embodied as a three-dimensional ecological path, offering a variety of landscape four seasons a year. The deck extended from the stroll path functions as a buffer zone between work area on the upper level and the civil service area on the lower level, and provides a pleasant view of Geumgang and the central green zone to citizens at the same time. The plane was planned with flexibility for future extension or renovation works. In particular, an appropriate position of the grand auditorium and the core was reflected in the plan to use the 2nd floor as a public health center and the 4th floor as a government building for the national assembly.

The mass and unique elevation, which was generated through the repetition of a curtain wall module for the reduction of energy consumption, made the project not as a towering authoritarian building but as a friendly and memorable landmark and an icon of the multifunctional administrative city.

순차적 건립계획

순차적 건립계획에 맞추어 1단계에서 청사, 보건소, 의회동이 통합운영시 공간의 활용성 및 효율성을 높이는 시스템을 완성할 가장 중요하게 검토되었다. 또한 2, 3 차 분이 완공되기 전 녹지공간의 활용에 대해서도 같은 고찰을 한 결과 청사, 보건소, 의회를 잇는 대축을 만드는 것이 가장 효율적인 대안으로 검토되었다. 이 대축은 야외 집회장으로 역할하게 되며, 3차분 완공시가 중축시 플랫폼으로 이용된다. 이 연속적인 대축는 청사의 모든 시설을 서로 연결시켜 주주 통행의 편의성을 증대시키게 된다.

1단계 2단계 3단계

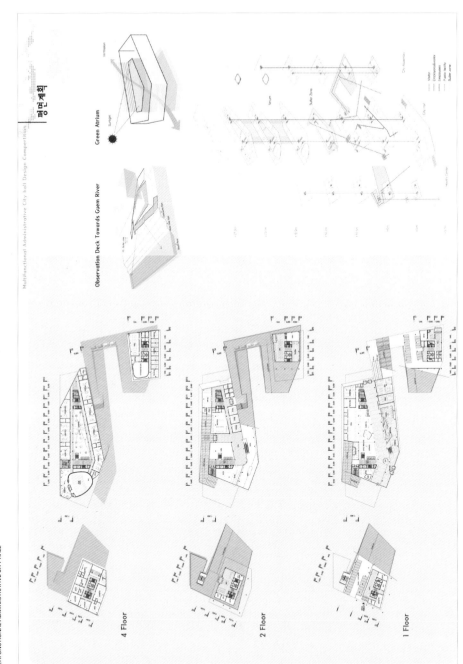

입·단면계획

입면도
Ventilation

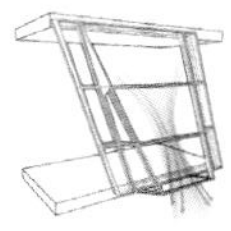

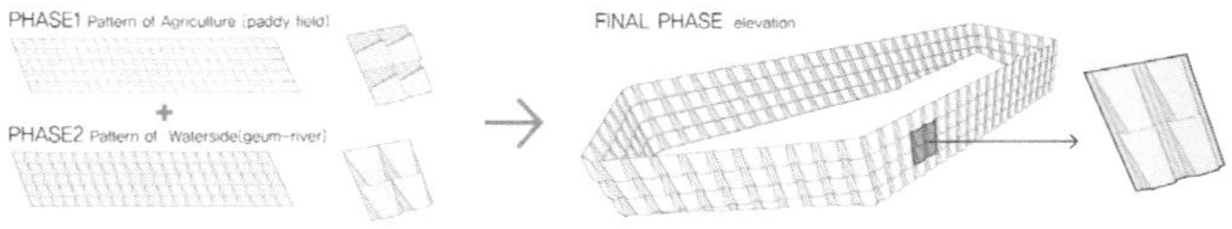

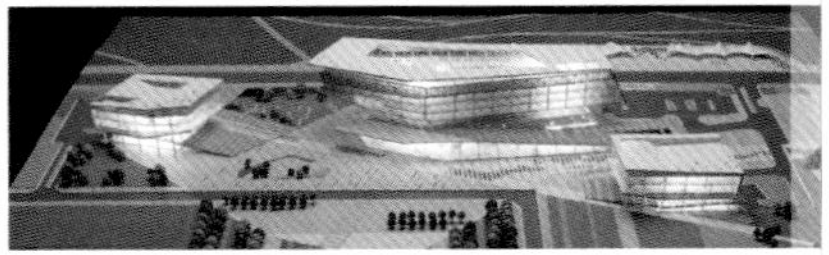

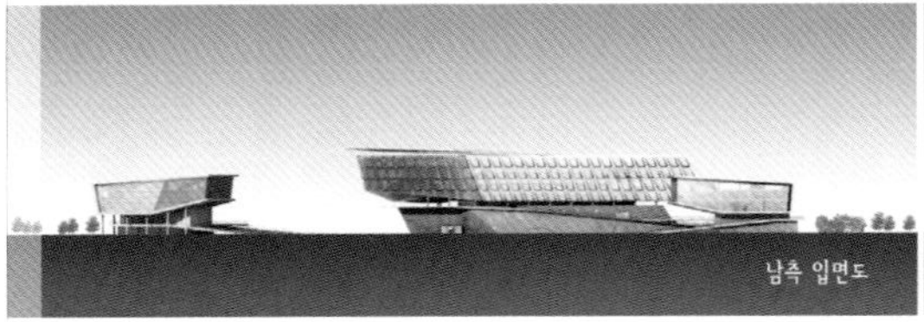

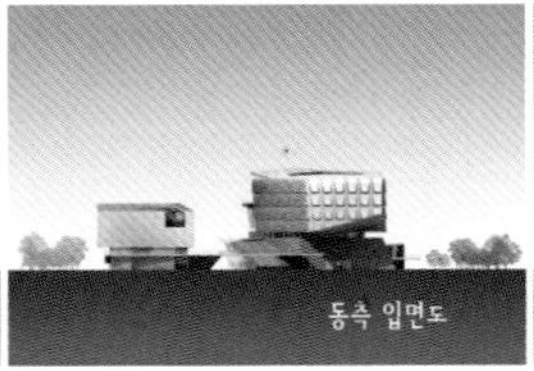

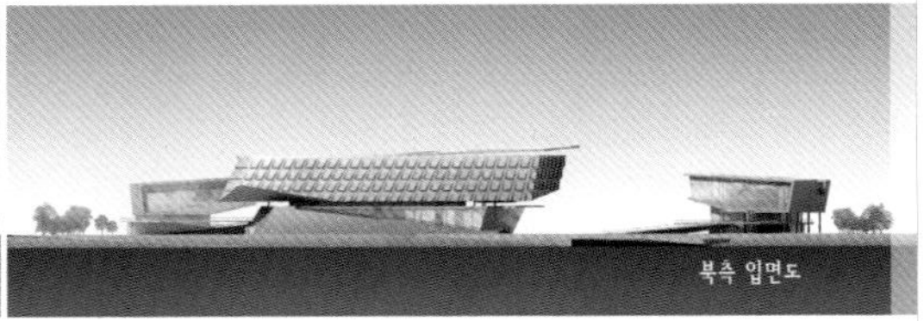

FINAL PHASE elevation

PHASE1 Pattern of Agriculture (paddy field)
+
PHASE2 Pattern of Waterside (geum river)

북측 입면도

동측 입면도

남측 입면도

단면간

종단면도
Oneway

횡단면도
Canal

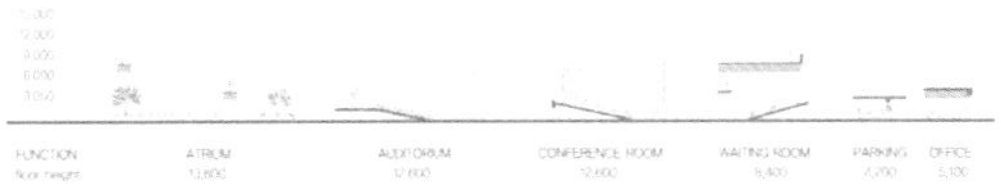

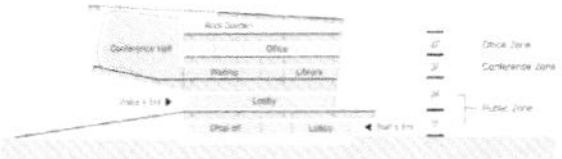

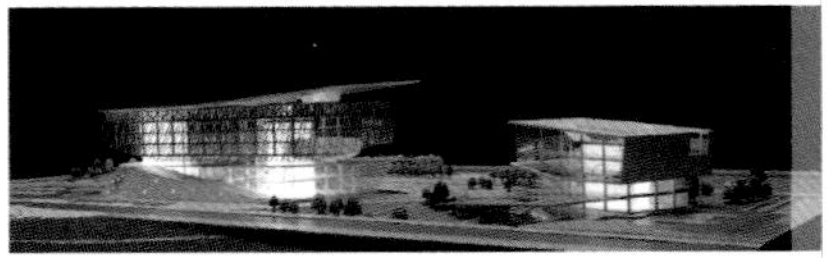

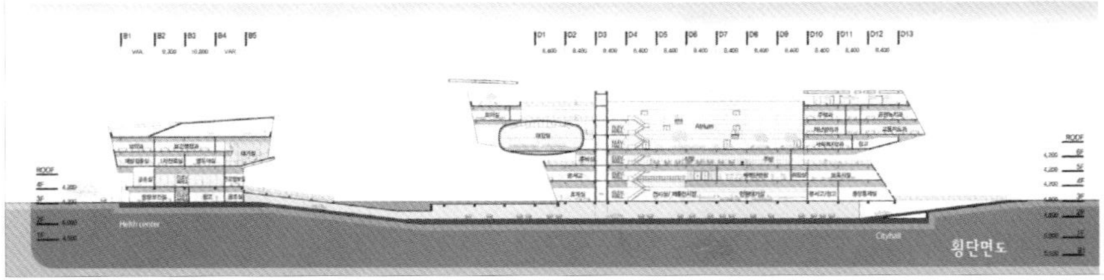

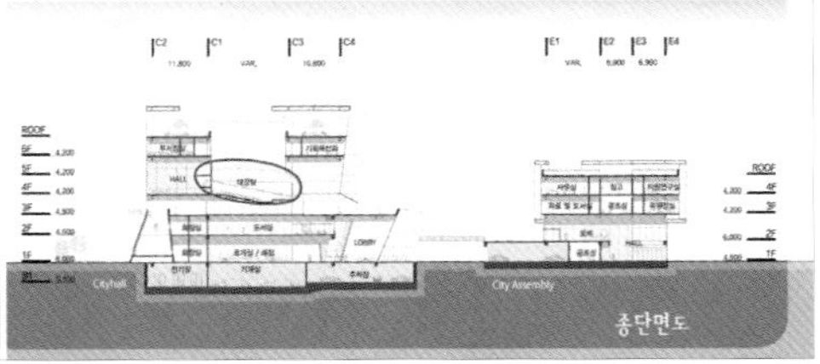

\>\> OFFICE

MULTIFUNCTIONAL ADMINISTRATIVE CITY HALL

Location Hotan-ri, Geumnam-myeon, Yeongi-gun, Chungcheongnam-do, Korea **Site Area** 41,661.00m² **Building Area** 10,609.38m² **Total Floor Area** 34,084.52m² **Landscape Area** 14,997.52m² **Building Coverage Ratio** 25.47% **Floor Area Ratio** 68.06% **Building Scope** B1–8F **Structure** Steel **Exterior Finish** Dabbed Granite, Pair Glass **Architecture Design** Kim Jong-cheon | KIAHAN Architecture **Design Team** Kim In-duk, Ko Jae-kyung, Jo Sung-rae, An Min-ju, Kim Young-min, Park Jung-su, Kim Wan **Client** Multifunctional Administrative City Construction Agency

CONCEPT Sejong city is a ring-shaped urban structure with a 'green heart' on the center. We propose 'Vertical Sejong' by designing a ring-shaped urban structure to express the unique urban icon of the city in the city hall. And through this project we realize a new icon for a city hall.

STRATEGY
- Urban Void: Multifunctional administrative city is a ring-shaped structure with 'Urban Void' on the center of it. It is an open urban structure without hierarchy and boundary. Focusing on an urban structure with a shape of a ring, we combine the vertical void inside the building and the horizontal void of the plaza to create a new public arbitral space of the city.
- Public Activity & Service: The city hall is a public service space. We propose 'Two city halls' which is divided according to the spatial purpose.
- Program: Through the administration-centered spatial composition and the characters that differentiate it from existing city halls, we design an open and cultural space and create a lively mass of programs that can draw out an interaction between the public and the city hall.
- Public Space (Communication): The green hub within the city is an 'Urban Void' for the city. Adopting a floating garden as the metaphor, it emerges from a monumental feature of the existing city halls to generate a new community space.
- Garden: The three gardens (inner garden, floating garden and base garden) with different characters function as mediators among the building, surrounding environments and the general public.

DESIGN PROCESS The 'Vertical Sejong' is a strategy that reflects the idea sought after by the multifunctional administrative city, separates a working space from a public space and maximizes the publicity. We introduced the urban structure and the flowing curve of Geumgang(River) in realizing it and focused on the sense of direction and motion, which reflect the mass in the layout of buildings. The buildings are positioned corresponding to the public programs on either sides of the city hall, which harmonizes with other facilities through the programs considering the phased development of a public health center and municipal assembly.

SITE PLAN The maximized publicity creates a space where various communities can be formed, not one that only satisfies administrative functions. We propose a layout of a city hall considering the phased development of a public health center and municipal assembly and the relation with the surrounding government buildings and by reinterpreting Sejong the multifunctional administrative city. The site in the vicinity of a green zone with a view of a landscape space is a public space used as a community space. As a mediator that links government buildings and nature and contrasted by the dynamic shape of the river, the green axis stretching afar breaks away from a single direction flow and adopts a sense of location corresponding to the horizontal force of the surroundings. This provides an open space embracing the entire city, melting the boundaries.

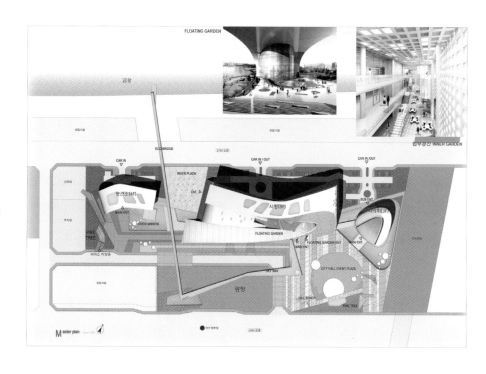

Site condition

시청사 청사의 도시구조적 특성을 나누어 분석 방법이 이용되어 있다가 하는 공간적 특성을 이용하여 다양한 공간을 이용하여 공간을 이용하여 다양한 공간을 이용하여 다양한 공간을 이용하여 다양한 공간을 이용하여 다양한 공간을 이용

NATURE & TERRITORIAL

이 건물은 도 시의 녹지스케치를 방해하지 않고서 방안을 제시해 보이고자 한다.
- 녹지거점 : Green hub
- 소거래 거리를 유기 기대효과를 기대하는 공간이 주기

GEUM RIVER / River, Flow, Fluid

이름의 흐름이 유동성을 가지는 자연요소를 건축적 요소로 환산해

| Green Network | Land & Water | Land & Open Space | Land & Urban |
| Commercial | Administration | Green | Education |

TERRITORY — SPACE
PLAN — URBAN
SPACE — VOID
EXPANSION IN & OUT — PUBLIC SPACE

Water Course — Green Zone — Velocity — AXIS
FLUX — SPACE — NODE — PROGRAM

Strategy

CITY HALL - Urban Void
우리는 청사형 도시구조적 주변하는 공간이 도시의 중심이 되어 다양한 공간의 중심이 되어 다양한 공간

A RING SHAPE
Urban — Urban Void
Fluid — Green Hub

CITY HALL - Public Activity & Service
청사의 공공적 성격에 맞추어 주변 시민의 요구에 맞는 공간적 공공성의 활동 공간을 계획하였다.
- 다양한 그룹 - 시민소통 공간

CITY HALL - Public Space(Communication)
도시속의 소통공간(Green Hub)을 주변 시민 Urban Void
공간의 연계를 Gathering Space로 활용 시민의 기본적 인
공동체적 문화적 삶(Communication Space)을 만들고자 한다.

OFFICE / WORK SPACE — EXHIBITION / CULTURAL — RECEPTION
FLOATING GARDEN

CITY HALL - Garden (Green Zone)
시청 건물 옥상고공의 정원 (Garden)과 옥상 정원공간을 공공에 개방하여 시민들과 함께 호흡하고자 한다. 시민에게 정원을 돌려주고자 한다.(Green Zone)

OFFICE / CONFERENCE — ADMINISTRATION / MEETING / REPOSE
GALLERY / SERVICE — EXHIBITION / GARDEN / AUDITORIUM / CAFETERIA

INNER GARDEN
FLOATING GARDEN
BASE GARDEN

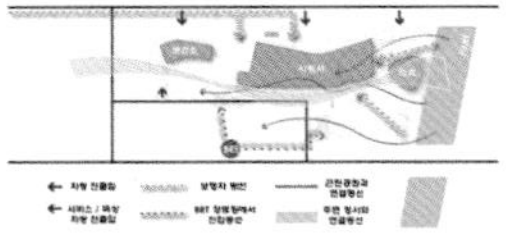

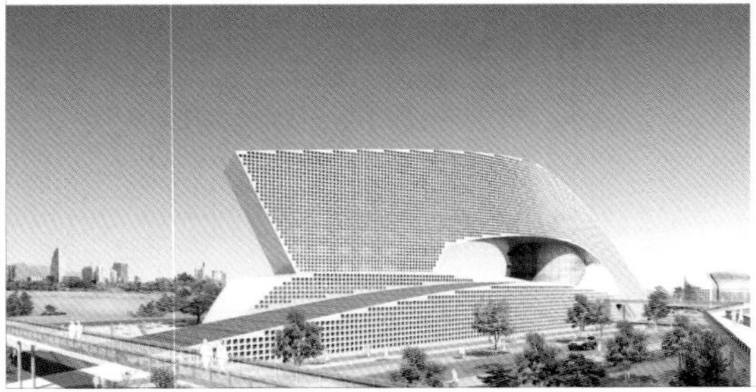

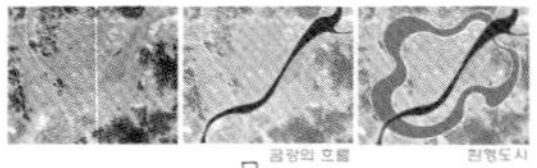

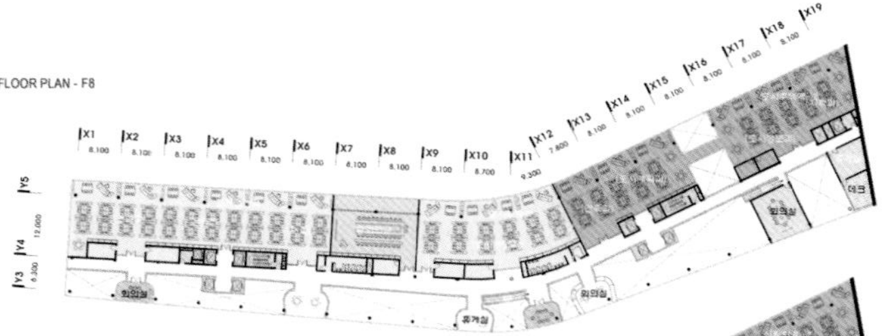

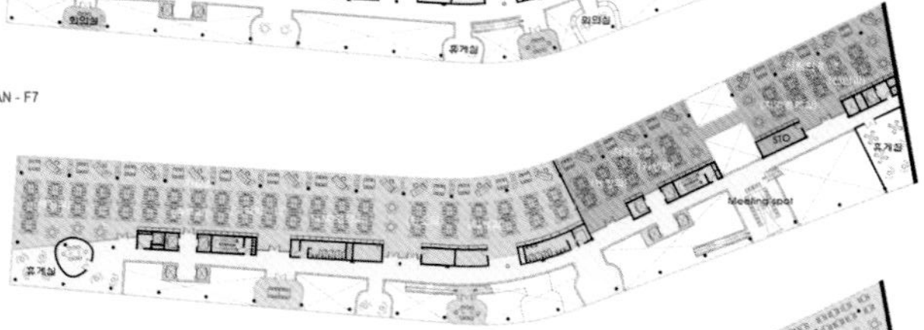

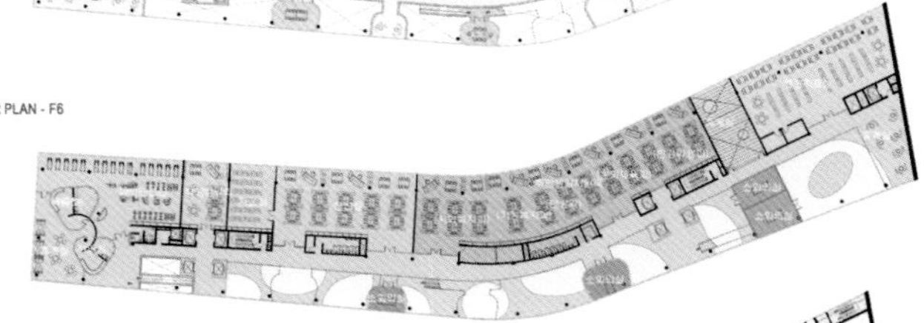

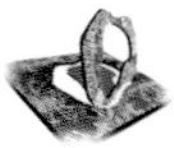

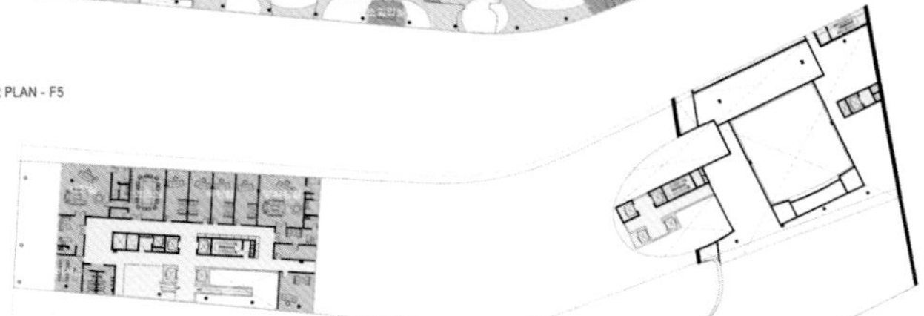

Design process

ICON - 행복도시의 상징적 이미지

VERTICAL "SEJONG"

Vertical Sejong은 행복도시가 추구하고자 하는 이념을 반영하고 직무공간과 공공 공간을 분리하며 공공성을 극대화 시키기 위한 전략이다
도시의 구조와 금강의 유려한 곡선의 이미지로부터 Vertical Sejong 실현시킨다

VIEW

우리의 시각은 방향과 움직임을 포함하고있다
이것은 건물의 배치에서 MASS를 변형하는 힘으로 작용한다.

CONNECTION

시청사의 좌우에 존재하는 공공 프로그램에 대응한다

Function Axis 대응하여 건물 배치

시청사를 둘러싼 외부 OPEN SPACE를 연결한다

Floor plan

FLOOR PLAN - F8

FLOOR PLAN - F7

FLOOR PLAN - F6

FLOOR PLAN - F5

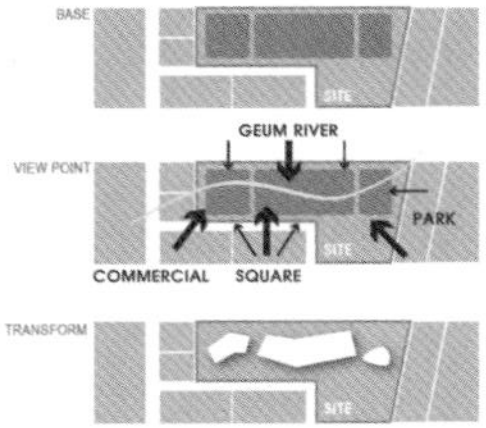

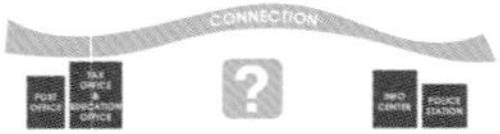

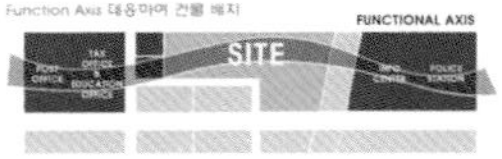

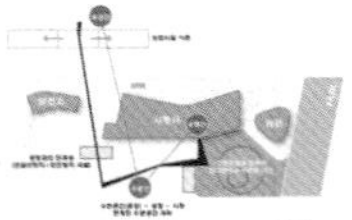

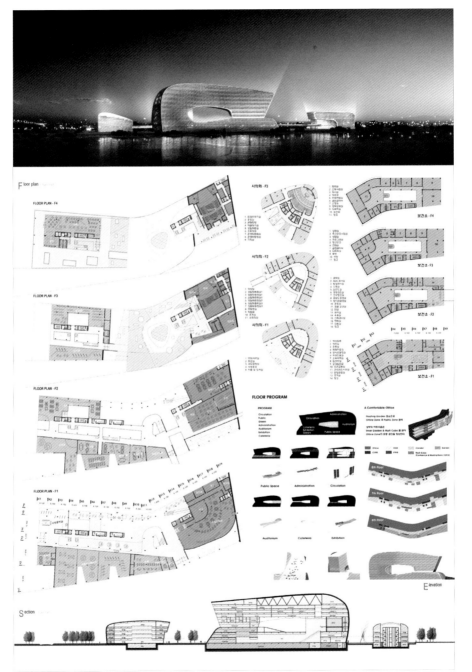

>> OFFICE

MULTIFUNCTIONAL ADMINISTRATIVE CITY HALL

Location Hotan-ri, Geumnam-myeon, Yeongi-gun, Chungcheongnam-do, Korea Site Area 41,661.00m² Building Area 13,970.02m² Total Floor Area 34,047.08m² Landscape Area 19,963.76m² Building Coverage Ratio 33.53% Floor Area Ratio 62.21% Building Scope B1~7F Maximum Height 35.4m Structure SRC, Steel Parking 226 Cars Exterior Finish Glass Curtain Wall, Aluminum Sheet, Wood Panel Architecture Design Kim Jong-nam, Ham In-sun | KUNWON Architects Planners Engineers Design Team Son Se-hyeong, Kim Jae-seok, Sin Dong-won, Kim Hong-il, Park Yeong-min, Kim Yeong-gyun, Jeong Ji-yeong, Lee Jeong-jin

LIVING FOREST / THE FOREST OF DAILY LIFE HOLDS THE CITY HALL The forest of the city hall is for the citizens. The city hall lies in the green shadow that comes into our daily life and memories with the pulse of the city.

FOREST FOR DAILY LIFE / THE CITY HALL HOLDING THE DAILY LIFE OF CITIZENS The plaza in front of the city hall is now represented as an urban daily space. It is a shelter for the rain and a resting area with a green shadow and a fine view of Geumgang(River). Such an image is achieved not by a large and static open space but by a kinetic space with active functions.

- City + Hall in Life / The vessel filled with the daily life of the citizens in multifunctional administrative city: The multifunctional administrative city hall involves programs that draw various circulation paths. The administrative street linking the city hall on the east-west direction is the central street that integrates a group of programs as the backbone connecting the administrative and public functions of the city hall.

- Stage of Nature / Urban cultivation land: The idle lands created from the phased development can be made into weekend farms for public programs of joint cultivation and harvest, and use the earnings from leasing the surface right of the partial land to cultural and public projects. This enables the citizens to experience the city hall more closely, and in the long run, functions as the foundation of a cultivation land of urban culture evolving into programs such as a night school, villa house for the aged and a citizen's exhibition center.

- Scene of Happiness / A place makes memories, and the memories dwell in the place: The interior and exterior public spaces generated from green roofs play not only the role of a city hall but also places for relaxation, gatherings, events and information acquisition, transforming the daily life of visitors into happy memories.

BREATHING FOREST / A CITY HALL AS A PART OF NATURE A forest is a space which citizens long to be close to, the energy of a city that provides a green air and shadow. As a mediator unlike the central green zone, the proposed forest of the city hall is an urban space with the function of communication.

- Shelter for Citizen / A hot spot in the city: The primary and central function of the living forest is to modernize the community spaces, which are being alienated from the city.

- Generator of Energy / The generation of environment-friendly energy: By installing solar panels on the roof, the multifunctional administrative city hall takes a step forward to the future-centered goal of the city: energy neutrality. We propose a vision as a good example of an energy self-sufficient city through the reduction of thermal load, reuse of rain water and the use of the materials with high water permeability.

- Unify Nature & Public / The forest created by time, nature and human being: The roof realized by nature and high technology play the role of a human shelter and a natural shelter.

GROWING FOREST / ORGANICALLY GROWING CITY HALL The city hall is an urban space. It stands on the extension of a decentralized development of the city. Although being government buildings, the three buildings classified by function constitute an organic aggregate to hold the contemporary and future contents and to create a dream society demanded by the city.

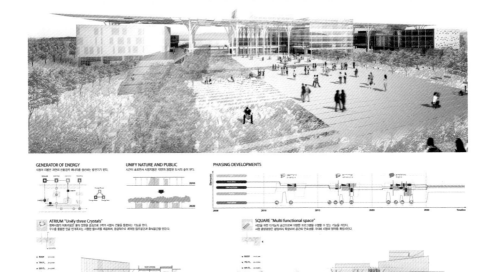

유기적으로 성장하는 시청

Growing Forest

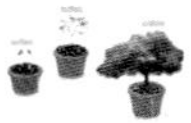

시간이 흐를 수록 도시는 더욱 풍부하게, 시간이 흐를 수록 시청은 시대를 앞서 빠르게, 시간이 흐를 수록 시민은 일상이 행복하게 행복도시는 유기적 성장을 모토로 삼고있는 성장의 도시이다.
행복시청사는 도시의 성장과 시민의 성장이 반응하여 성장한다.

GROWTH OF THE CITY / 도시의 성장

행복도시는 자연의 성숙과 더불어 만들어지며, 마치 숲과 같은 유기체적 도시로써 아이덴티티를 가진다.

GROWTH OF THE CITY + HALL / 시청의 성장

시청과 보건소, 시의회의 기능이 단계별로 분동형 개발이 되는 행복시청은 단계별로 기능의 확장과 도서실, 전시실, 생활편의시설등 풍부한 시민 프로그램들의 도입을 통하여 시간의 흐름에 따라 문화적인 성장을 하는 진정한 도시공간(City + Hall)이 된다.

GROWTH OF THE CITIZENS / 시민의 성장

성장하는 시청과 함께 호흡하는 시민들은 일상적 참여, 기능적 참여, 자발적 참여를 통해 점진적으로 다양한 문화를 체험하게 되며, 서로간의 커뮤니티를 통하여 문화적으로 성장하게 된다.

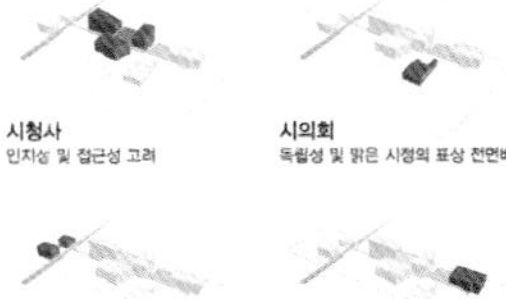

시청사
인지성 및 접근성 고려

시의회
독립성 및 맑은 시정의 표상 전면배치

보건소
시민건강 중심센터

대강당
시민들을 위한 도시문화 프로그램

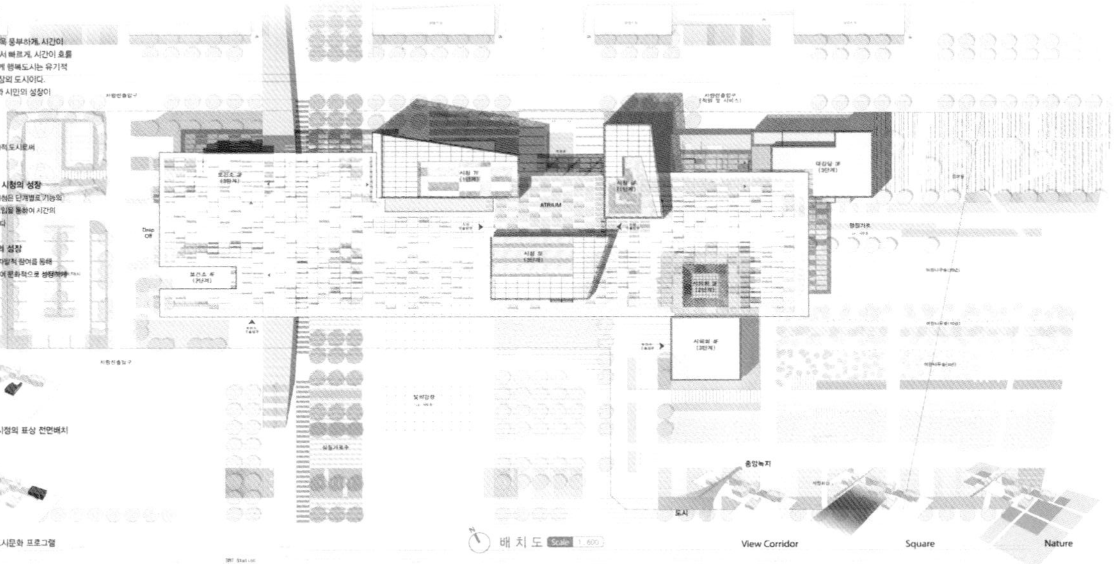

배 치 도 Scale 1:600

Three Elements in Forest

자연에서 솟아오른 3개의 수정체 형상을 모티브로 하여 시청사의 Mass를 구성하였다.
3개 시청의 입면은 금강을 향한 조망축을 열어주는 의미에서 투명한 유리를 사용하였다.

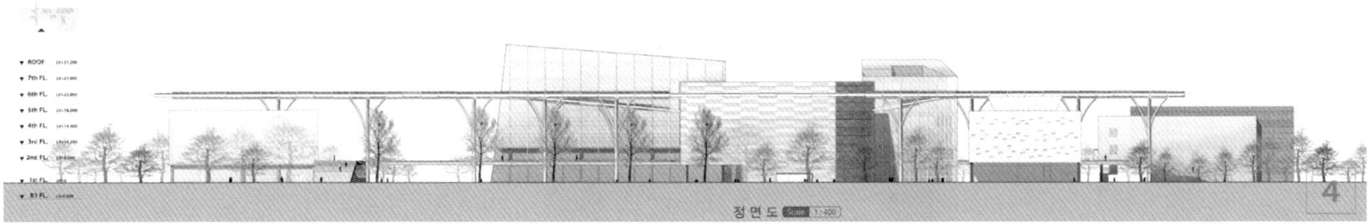

정 면 도 Scale 1:400

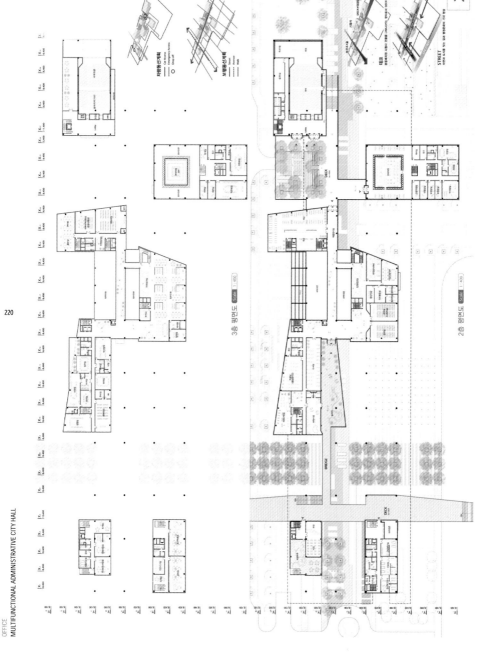

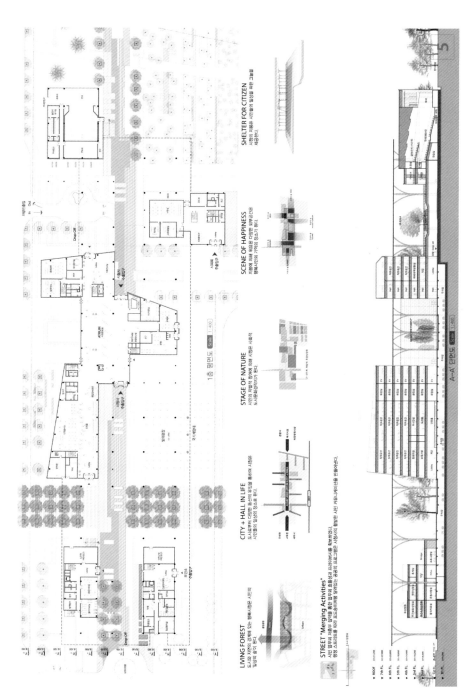

\>> OFFICE

THE PROVINCIAL OFFICE & ASSEMBLY OF CHUNGCHEONGNAM-DO

Location Hongbuk-myeon, Hongseong-gun, Chungcheongnam-do, Korea & Sapgyo-eup, Yesan-gun, Chungcheongnam-do, Korea Site Area 334,000m² Building Area 94,785.84m² Total Floor Area 104,761.15m² Landscape Area 239,214.16m² Building Coverage Ratio 28.38% Floor Area Ratio 21.31% Building Scope Provincial Government_B1~7F / Council Office_B1~3F / Fire Department_B1~2F Structure RC, Steel Main Finish Color Pair Glass, Metal Panel, Concentrating Solar Panel Architecture Design Yi Gwan-pyo, Kim Young-chan | AUM & LEE Architects Associates Design Team Sim Jae-kun, Jo Il-hwang, Du Gwang-ho, Im Sang-do, Ma Sun-seok, Jeong A-Yeong, Kim Jin-u, Lee Jin-yeong, Jo Im-seon

RHIZOME: PLURALISM(DEMOCRACY) The configuration of the provincial transfer site stands for the pluralism to intentionally disperse a program and a division of the space. The rigid bureaucracy and the right are sublated and a horizontal equal concept to respect an individual value is projected. Therefore, the master plan of the new administrative site is unified as an independent tangible and intangible buildings keep an equal relationship, are dispersed closely and holds a natural boundary with the topological environment.
Multi-Characteristic, Organic Connection, Communication, Interaction

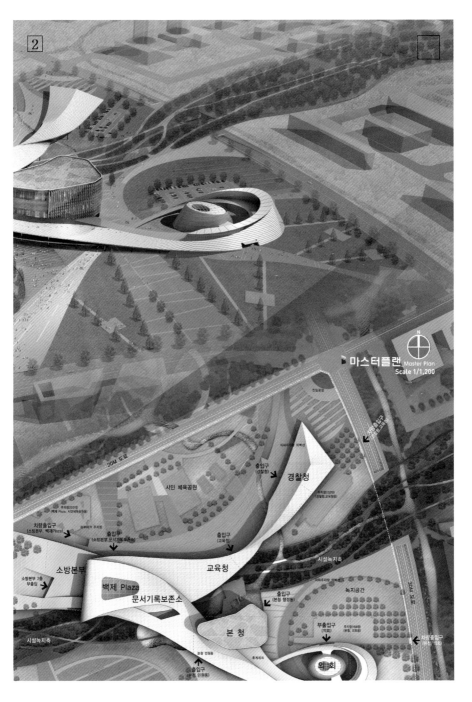

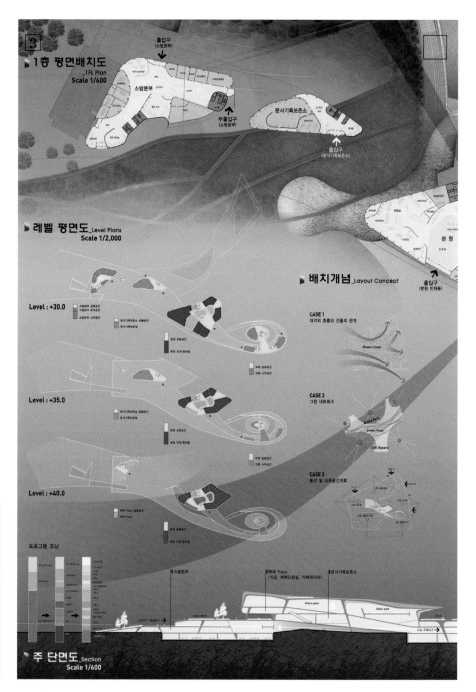

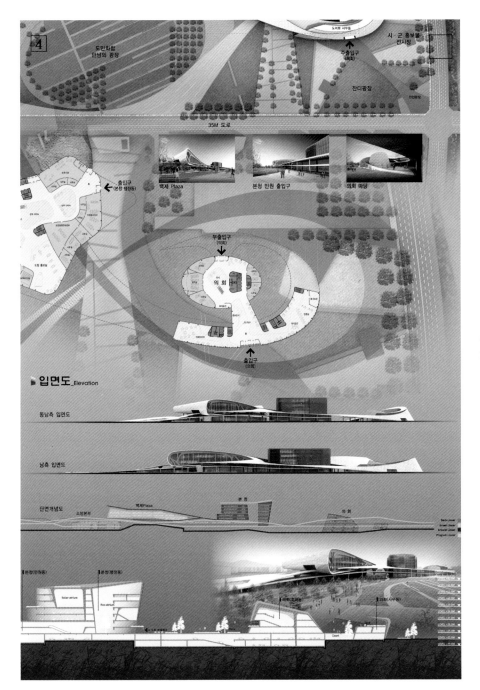

\>> OFFICE

NATIONAL INSTITUTE OF MARINE BIOLOGICAL RESOURCES

Location Janghang-eup, Seocheon-gun, Chungcheongnam-do, Korea Site Area 132,011.00m² Building Area 21,539.95m² Total Floor Area 36,350.75m² Landscape Area 28,382.36m² (21.50%) Building Coverage Ratio 16.32% Floor Area Ratio 27.54% Building Scope B1-4F Maximum Height 30.6m Structure SC, Truss, SRC Main Finish Aluminum Punching Metal, T3 Aluminum Sheet, T24 Low-E Pair Glass, Aluminum Composite Panel Parking 180 Cars Architecture Design Jeong Young-kyoon, Kum Doo-yeon | HEERIM Architects & Planners + Kim Yong-gwon | DOSHIIN Architecture Design Team HEERIM_Kim Jin-su, Kim Dong-hun, Ruy Mu-yeol, Lee In-su, Kim Hae-jin, Jeong Yong-ju, Kim Young-tae, Lee Sang-heon, Yeo Jun-su, Yoo Ji-sang, Jordan Trachtenberg, An Ji-eun / DOSHIIN_Lee Chang-yeon, Kim Su-jeong, Jeon Hyeon-jeong, Kang Jeong-hui, Lee Geun-taek, Kim Su-i-a, Yun Kyung-hyeon, Kim Se-young, Jeong Jeong-a, Jang Hui-jin

PROJECT BACKGROUND National Institute of Marine Biological Resources studies the classification and system of marine biological diversity at the global level and publicizes the achievements, and also functions as a resource bank that provides the source in relation to marine biotechnology (MBT). Thus, the project should be the center of both domestic and international conferences, and the research and educational organization for marine biological resources. We also considered the connectivity with surrounding natural sightseeing destinations for the purpose of providing culture and relaxation spaces for local community as a recourse center participated by local residents and visitors. The environmental space linked with the beautiful surroundings of ecological mud flat and habitats seeks for creating organic architecture that coexists and lives together with nature and symbolizes the role of National Institute of Marine Biological Resources in the field of natural science.

DESIGN IDEA The ocean communicates through the rising flow of water, and a new life form is born from such rich resources. The land ceaselessly communicates with the ocean through the ebb and flow. The natural flow of the west sea ascends and is realized in the land. In the flow of upwelling, National Institute of Marine Biological Resources will become a place of communication where human beings and nature coexist.

DESIGN CONCEPT The horizontal and vertical flow of mass reflects the interconnection between the ocean and land through the ebb and flow, and the creation of nature through flowing water. The main concept of the design is 'Blue Spiral', a compound made of 'blue' that symbolizes the ocean and 'spiral' that signifies the dynamism of water and the circulation flow of the exhibition.

The facilities for research, education, exhibition and collection were centralized around the aquarium, which was planned to be built in the south, thus shortening the circulation of researchers at the maximum level. And making the best of the natural topography in the vicinity, we created an optimum research environment by taking into consideration the rooftop area, view, lighting and ventilation for researchers. The deck was planned so that the circulation patterns for visitors and researchers wouldn't overlap, and the exhibition and education facilities were located on the front of the site to establish a three-dimensional circulation system.

The shape of the mass expresses the vertical movements of the wave and features the horizontal and vertical linear exterior and the penetrating outdoor spaces. As for the elevation, the designer applied the frames inferred by the unit fabric of a cell and the scale patterns of fish to a punching metal, which shows a rhythmical gradation of the ebb and the flow of the tides to unify the concept of the oceanic life.

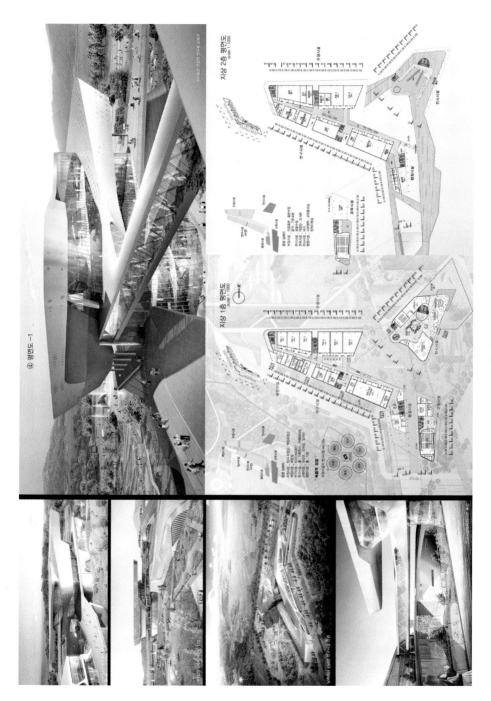

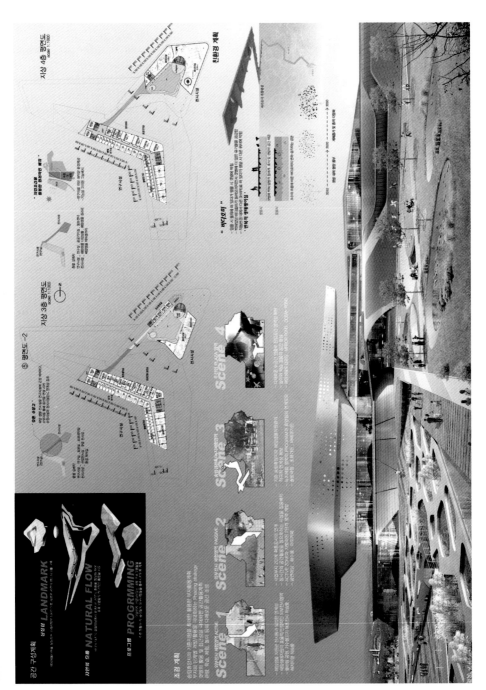

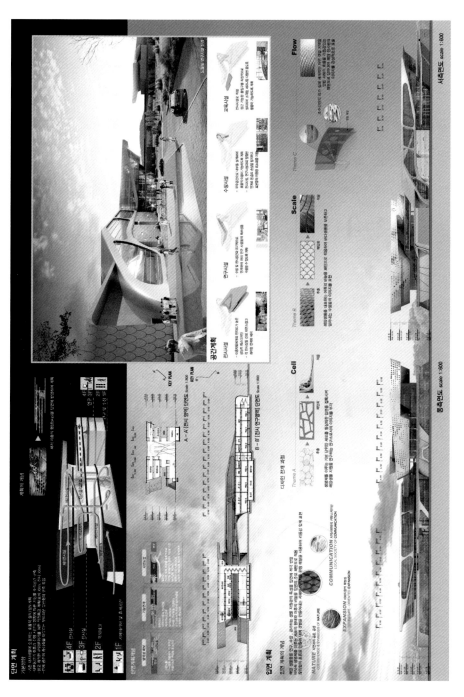

\>\> OFFICE

GIMPO-YANGCHON ECO CENTER

Location Unyang-dong, Gimpo-si, Gyeonggi-do, Korea Site Attribute Semi Resident Site Area 9,993m² Building Area 1,862.07m² Total Floor Area 3,437.51m² Landscape Area 6,062.57m² (60.67%) Building Coverage Ratio 18.63% Floor Area Ratio 19.31% Building Scope B1~3F Structure Steel, RC Parking 47 Cars Architecture Design Kim Seon-jae | YEZU Architects Planners & Associates

This Eco Center is put emphasis on the changed building to carry out a role for the PR booth of the Land & Housing Corporation as a land mark to inform the Gimpo Eco resident complex and an office of the Task Force and then to be remodeled as an Eco-Center to observe and experience an aqua ecology and a habitat for migratory birds of the brackish water zone in the Gimpo plains and to be a gateway of an Eco-Tour. The region of Yang-gok, Gimpo is the biggest habitat for the winter migratory birds in the Han-riverside, as the Han river and a waterway swamp plain meet, raises an ecological value of an existing habitat for migratory birds as a plan for restore a half-moon swamp and after making a Gimpo Eco resident complex, shall carry out a role for a place to learn an ecology which nature and human live together.

BLOCK PLANNING
- Plan for a building to make a shape for an uplifted stratum which a mass of the Land Form to utilize a shape of an existing land rise toward the Han river from a square and make a minimal influence into a neighboring ecosystem
- Make a step-type outdoor space from the roof garden to the wetland ecology to utilize a roof-floor of the Eco Center, as a land level plan (a wetland ecology, a grassland ecology and a waterway ecology) to match for the half moon type swamp in the northern side of the land

PLANNING
- Two stems of mass to divide into a PR booth, a northern side and an office of the Task Force, a southern side are linked through a underground space and form a circulation movement centered at the lobby
- A sphere of an office of the Task Force in the southern side shall be remodeled as an exhibition facility afterward and forms two exhibition sphere around the lobby

SECTION PLAN
- Induce a visual experience to rise and change as rising along lamps from spheres for an education, a management and an accessory in the basement following a shape of an uplifting building and reaching an exhibition sphere.
- An observatory to view the ecosystem around the Han river and the new Gimpo city to shall be planned for a gradient tower to match a shape of the building to rise and uplift, for utilizing a nature-friendly materials and for an eco-tower to cross an indoor and outdoor spaces repetitively.

ELEVATION PLAN
- Form an image which the land and the building become one by going after a shape of a stratum uplifted from the land and stacking up the blocks made of loess, an eco-friendly material horizontally.
- The edging part of the mass to meet the Han riverside shall be secured for a view of the Han river by a plan for a curtain wall

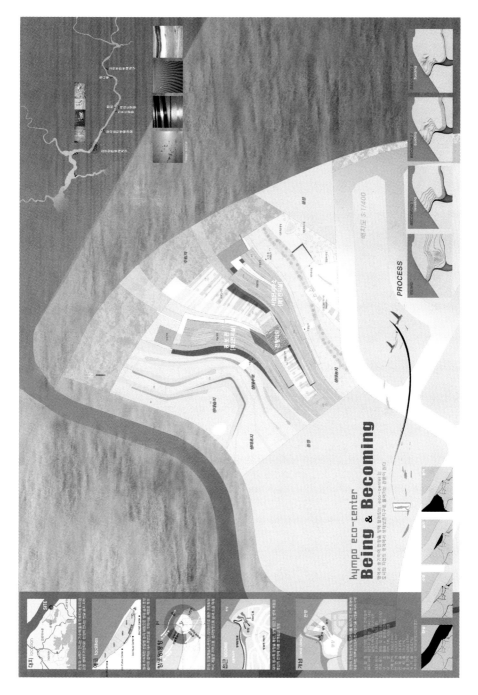

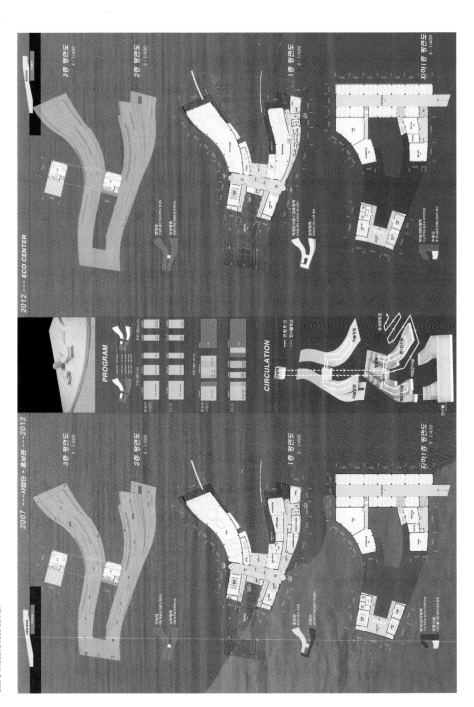

>> OFFICE

GIMPO-YANGCHON ECO CENTER

Location Unyang-dong, Gimpo-si, Gyeonggi-do, Korea **Site Attribute** Semi Resident **Site Area** 9,993m² **Building Area** 1,788.54m² **Total Floor Area** 3,562m² **Landscape Area** 1,231.65m² **Building Coverage Ratio** 17.89% **Floor Area Ratio** 28.17% **Building Scope** B1–3F **Structure** RC Rahmen HYBRID(High Performanced Con'c+Post-Tension, SRC) **Height** 30.4m **Parking** 38 Cars **Exterior Finish** T24 Pair Glass, Aluminium Composite Panel **Architecture Design** Kang Chul-hee | Idea Image Institute of Architects + Kim Ki-jong | PIVOTEC **Architecture Design Team** Kim Sang-jin, Park Yeon-seong, Mun Hui, Kim Jong-mu, Kim Bo-geon, Kim Byeong-jin, Kim No-a | Idea Image Institute of Architects **Exhibition Design Team** Jeon U-gong, Kim Seong-nam, Kim Jin-kyeong, Hwang Hyeong-sik | PIVOTEC

CONCEPT
· Gimpo Eco Center, an aggregate of eco-friendly factors: envelops a cuticle of the space as a portion of the land which becomes one with the nature. (Envelope with natural cuticle of land)
 - Earth of the land becomes a roof, is a piece of cuticle to protect the below space and a filter possible for human various actions.
· Energy Tower: Secure a sustainability of the building through a symbiosis with the nature and a use of eco-energy.
 - The cutting edge new recycling energy system of the solar energy necessary for the building by means of a convection through the solar heat.
 - The section to learn an ecology by means of the greenhouse effect of the solar heat
 - A view to watch the Han river and the new city, and an execution for a role of a land mark for the new Gimpo city
· Variable space not an artificial: vividness of change (Flexible space)
 - Easy for a plan for an eco-remodeling to pursuit a strength and weakness, a rhythm and visual variety of the space flow through a horizontal space of the office and a vertical slope space of the PR booth
 - The space can be widened or reduced by a type of exhibition and a flow by a column-absent space not to have any column within the floor and a design of variable wall
 - In a function, a flexibility is given.
MASS COMPOSITION PROCESS Eco-Flux with uncertainty principle space: A nature-friendly Eco Skin and a continuous exhibition space to be made along a close traffic line shall be the factors to harmoniously an intrusion and an expansion of the ecosystem and it is a indefinite space not to define by dividing a sectional program into sectors and will be a space to contain an indefinite behaviour more naturally.
FLOW PLANNING
 - After inducing naturally from the outside of the building, an inducement for a continuous watch into a vertical promenade through lamps
 - Because each inner space is independent on the function but they are linked spatially, it is planned to rouse various experience feeling through an opening toward multi-direction.
PLANNING
 - Sphere of Exhibition: It is composed of an autonomous selecting traffic flow to open various possibility without watching along a compulsory circulating movement within the exhibition space. Each factor to definite a traffic line of the space is formed for subsidiary facilities with various functions and the resting area and they are responsible for roles to help spectators solve a lot of factors necessary in the middle of an exhibition.
 - Sphere of Task: It utilizes an Office Landscape System so that it is easy to transform a space for task into a space for exhibition.

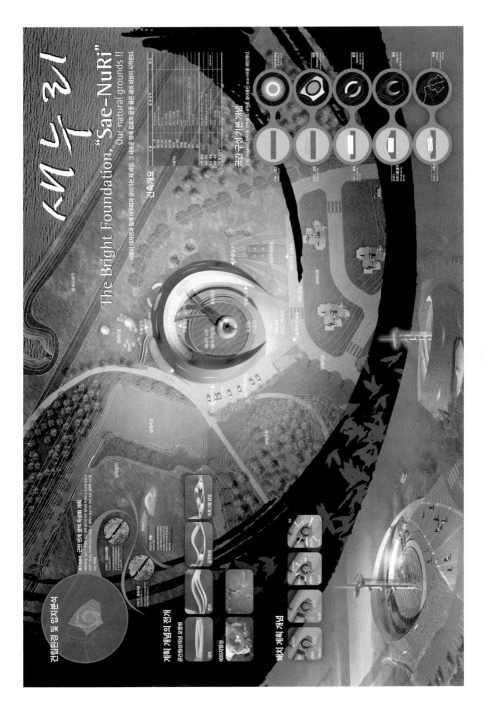

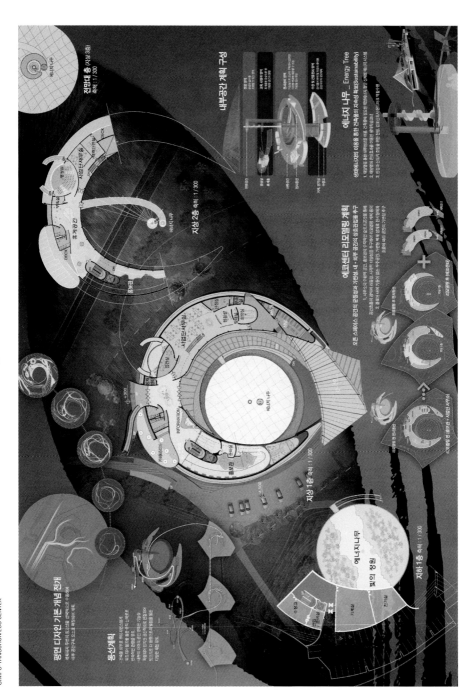

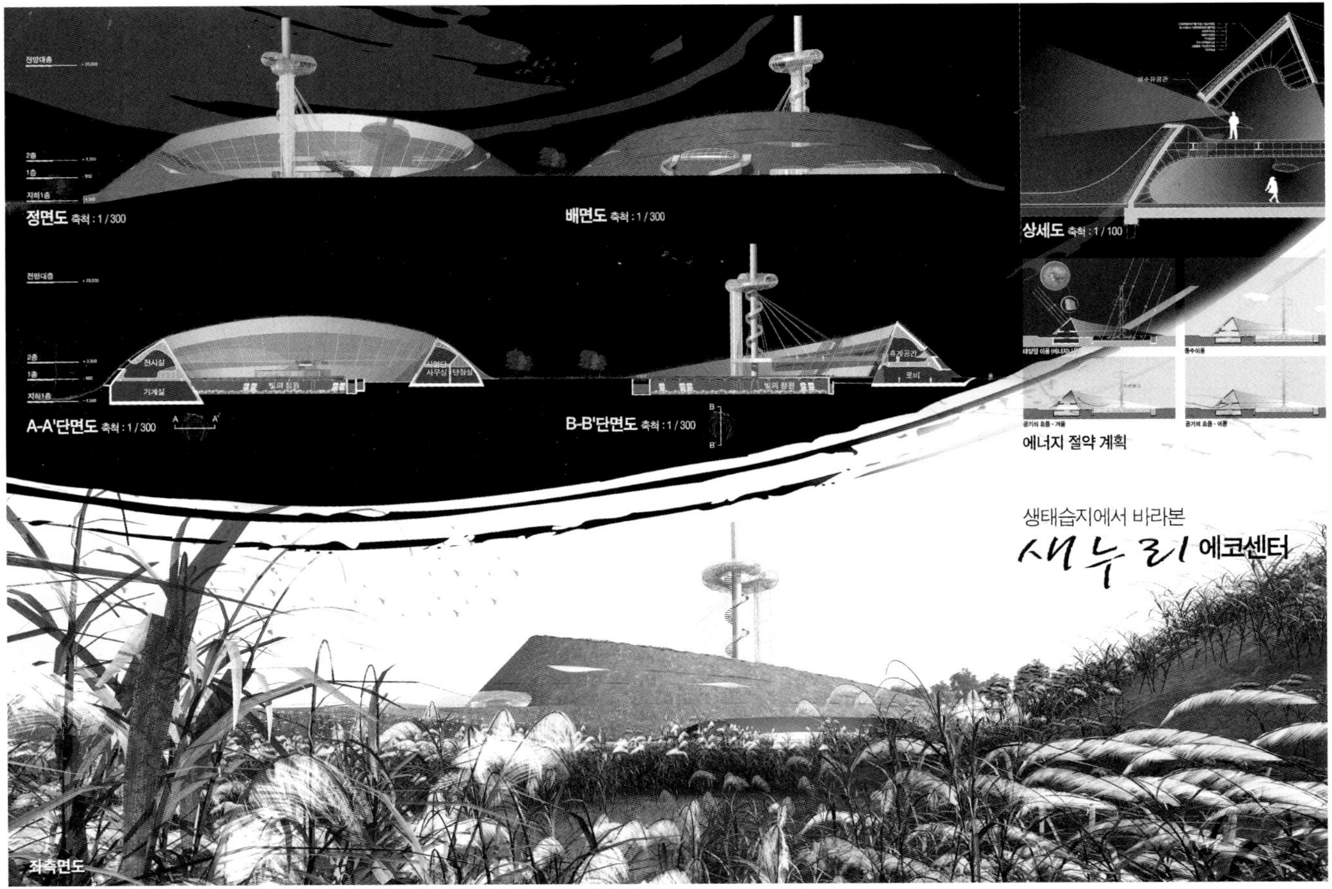
정면도 축척 : 1 / 300
배면도 축척 : 1 / 300
상세도 축척 : 1 / 100
A-A'단면도 축척 : 1 / 300
B-B'단면도 축척 : 1 / 300
전망대층
2층
1층
지하1층
전시실
기계실
빛의 정원
사무실
단장실
휴게공간
로비
에너지 절약 계획
생태습지에서 바라본
새누리 에코센터
좌측면도

\>\> OFFICE

GIMPO-YANGCHON ECO CENTER

Location Unyang-dong, Gimpo-si, Gyeonggi-do, Korea Site Attribute Semi Resident Site Area 9,993m² Building Area 2,408.05m² Total Floor Area 3,541.58m² Building Coverage Ratio 24.10% Floor Area Ratio 29.27% Building Scope B1~3F Structure RC, Steel Parking 61 Cars Exterior Finish Color Aluminium Sheet, T22 Pair Color Glass, Atmospheric Corrosion Resistant Steel Plate Interior Finish Wood Flooring, Stone Architecture Design Chung Soon-geun, Kim Won-ki | NODE Architecture Design, Inc. + Lee Han-gi | Daelim University Design Team Choi Won-jong, Bae Jeong-a, Yun Ha-yeon, Park Seong-hwan

CONCEPT OF DESIGN
ECO-SHELTER
The concept of Gimpo Eco Center starts a concept of a boundary of the city and the nature.
Complex and holy experience to the vegetation the river and swamp and pure water...
The concept of design puts an emphasis on the close form like a nest structure and the simple mass.
· Water - City of Water, Gimpo
 - Water in Gimpo is an absolute existence to guarantee an agriculture and a survival.
 - The concept of fusion of water and agriculture which is an important factor in the Korean traditional gardening is introduced.
 - Making a waster space to escape from the city life and feel a mental space.
 - Giving a maximum view to the northern waterside space.
· Shape - Nest
 - Induce a symbolic image of a nest structure composed of the weatherproof steel
 - Experience a space within a nest in an indoor space in daytime.
 - Express a shape of a nest filled with a light in night
 - Floating space composed freely within a nest space
· Sustainable Design - Geomancy
 - Arrangement to consider a wind direction in winter and summer
 - Plan of mass according to an analysis of the solar altitude
 - Effect to descend the neighboring temperature in summer by introducing a water space
 - Prevention of in-flow of sunlight with an outer cover of a nest structure in south

BLOCK PLANNING The Gimpo Eco Center is arranged combining new neighboring environment with another codes (water space, square, park) considering the movement of wind and sun as well as the existing environmental factors, Sat river, pasture, Han river in the front side.

PLANNING I accept all the plains of three stages(first stage : Task Force office + PR booth, second stage : eco center + PR booth, third stage: eco center) demanded in the ordering organization, and make an unlimited space for a plain possible to change. In the end, I borrow multi-purpose plain form without a collision of an expected program with the existing one and at the same time plan a Dual Space which a spatial wavelength of the PR booth, an exhibition occurs in the inner and outer sphere.

ELEVATION PLAN
· "Communication" - NEST: Make a projection of a light with mysterious feeling, not a daily life inside and outside of the architecture (symbolic expression with regard to the communication of the nature and the human and making various expressions of the architecture)
· "Communication" - floating mass: form rich and various feelings of the space
· Combination of mass: the floating mass and the structure with the nest shape interact effectively in absorbing a light in the space.

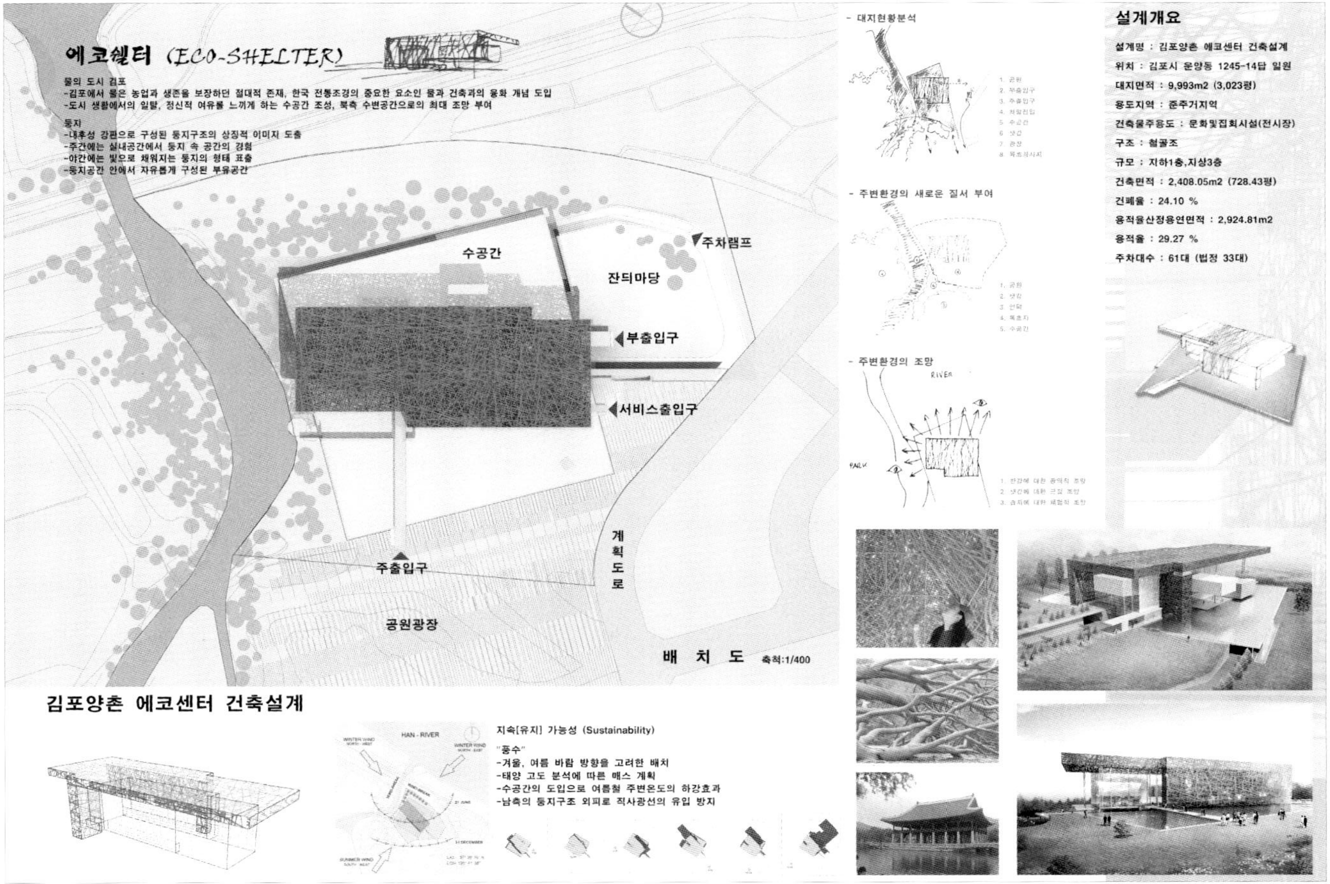

에코쉘터 (ECO-SHELTER)

통일 도시 김포
- 김포에서 많은 농업과 생산물 보관하던 법제적 온재, 현재 전통적인 중심인 소으신 곡자 건축과의 물화 개념 도출
- 도시 생태에서의 일물, 정난지 뉴앙을 느끼게 하여 수공간 조성, 계속 수변공간들의 최대 조명 부여

특징
- 내부공 김포공원 주민한 동자조직 성장의 이미지 도출
- 주민대 소제공간의 건축내 중심적 명료
- 아침사 대모로 혜의문 성장 형태 결여
- 동지공간 인해서 구조된 부출공간

설계개요

설계명 : 김포양촌 에코센터 건축설계
위치 : 김포시 양촌동 1245-14번지 일원
대지면적 : 9,993m2 (3,023평)
용도지역 : 준주거지역
건축물용도 : 문화및집회시설(전시장)
구조 : 철콘조
규모 : 지하1층,지상3층
건폐율 : 24.10 %
용적율산정용면적 : 2,924.81m2
용적률 : 29.27 %
주차대수 : 61대 (법정 33대)

- 대지환경분석

- 주변경관의 새로운 접시 부여

- 주변경관의 조망

배 치 도
축척:1/400

주차램프
건되마당
부출입구
수공간
서비스출입구
주출입구
공원광장
계획도로

지속(유지) 가능성 (Sustainability)

홍수 :
- 겨울, 여름 방향을 고려한 배치
- 대지 고도 분석에 대른 계획
- 공간고 도입으로 여름 주 편도도 하강효과
- 남북으 동서구조로 외피로 직사광선의 유입 방지

김포양촌 에코센터 건축설계

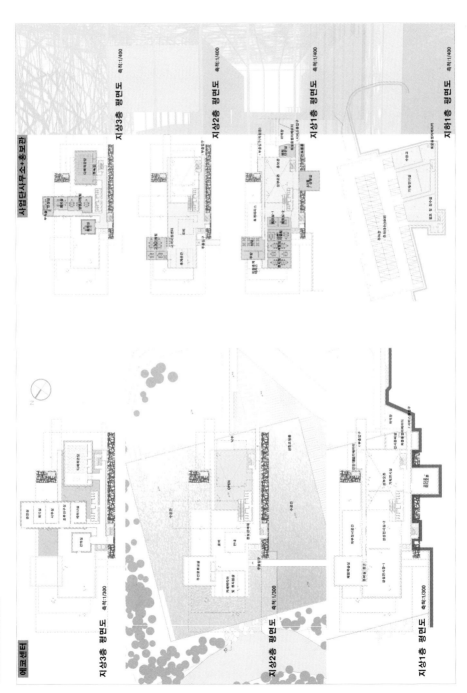

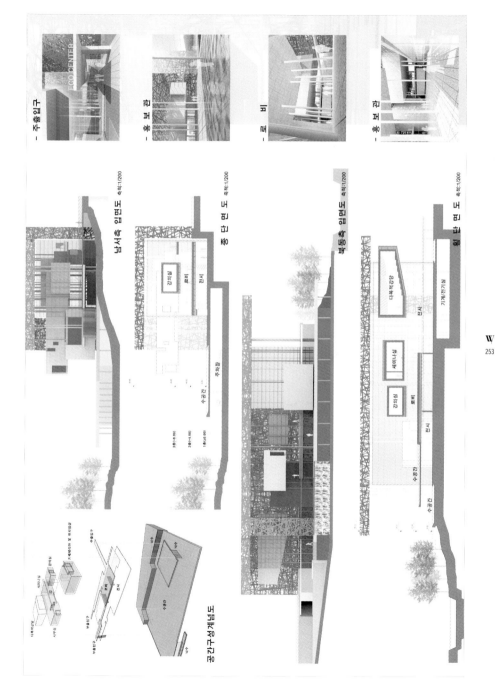

Interactive Networking Ground

>> OFFICE

COMPLEX-THEME PARK TOWN OF CHEONAN

Location Munhwa-dong, Cheonan-si, Chungcheongnam-do, Korea Site Attribute General Commerce Site Area 22,642.00m² Building Area 5,169.34m² Total Floor Area 28,772.13m² Building Coverage Ratio 22.95% Floor Area Ratio 94.26% Building Scope B2~12F Structure RC(Part Steel) Exterior Finish T24 Pair Glass, Aluminum Composite Panel Parking 211 Cars Architecture Design Yoon Se-han, Nam Ki-hong, Kim Tae-man | HAEAHN Architecture + Jo Yeong-don | USUN Engineering + Hwang Dong-sik | DAMOOL Architects Design Team HAEAHN_Im Jae-ho, Gwon Jin-seong, Park Hui-jun, Jeong Man-cheol, Gwak Geon-seop, Kim Kyeom-hyo, Lee In-ho, Yun Won, Jeong Won-yeong, Choi Ji-ho, Lee Ae-ran, Lee Se-yeong, Jo Sun-hui, Han Hye-rim / USUN_Bae Tae-yong, Hwang Ha-yong, Kim Ik-su

PLANNING STRATEGY Our strategy of a complex-theme park town is to promote the revitalization of the original urban center and make a themed landscape with an organic relation with architecture, thus creating a local destination as well as a cultural and education place for Cheonan citizens. As for specialized facilities, we aimed at establishing a children's center as a culture experience space with new technology and concept and at the same time an advanced multifunctional entertainment space, a youth training center as a sensuous relaxing space and lively activity space and a work facility as an environment-friendly human culture space.

Design Concept

· Program Analysis: The connection possibility of three programs with different characters is drawn out to structure a program network and create a new function that can be complemented.

· Planning Concept 1: Interactive Networking Ground of Pleasure

"A street that tells one's life from a child through an adolescent to an adult… a place where everyone regardless of generation harmonize and enjoy various activities… a courtyard in which people can make their own spaces… a jolly ground…"

· Planning Concept 2

- Network: The concept of connection, instead of combination or unification, ensures cultural contents that can be shared by the children's center, youth training center and the work facility at once.
- Community: The gatherings and unity of groups with diverse ages and attributes create a community plaza using the node among spaces apart from embodying functional spaces.
- Flexibility: To realize a variable space that transforms itself according to time and season.
- Identity: To ensure individual independence by analyzing the character of each group.

LAYOUT PLAN The accessibility is enhanced by the connection with the existing underground shopping malls linked to Cheonan Station. And the main pedestrian path encircling the entire site has an organic connection with the plaza and the green zone that adapts to the flow of nature. It will have a three-dimensional flow extends not only to the interior space of the group of facilities and the rooftop garden but also to the observation tower.

ELEVATION PLAN The mass of the lower part is actively harmonized with the green zones and the park as an extension of the park for intersection of the interior and exterior spaces. The middle part has a different human and urban scale through a dynamic elevation considering the view from the plaza 'The Bell of Citizens' and through an intense and simple elevation considering the velocity and view from the outer road.

SECTION PLAN A buffer zone was planned in the vertical zoning space between the facilities for work and children to create a separate outdoor space of the children's center and at the same time divide the function more distinctly, and also ensure a nature-friendly pleasantness through 'Green Cube', which is the atrium embracing the children's center and the youth training center.

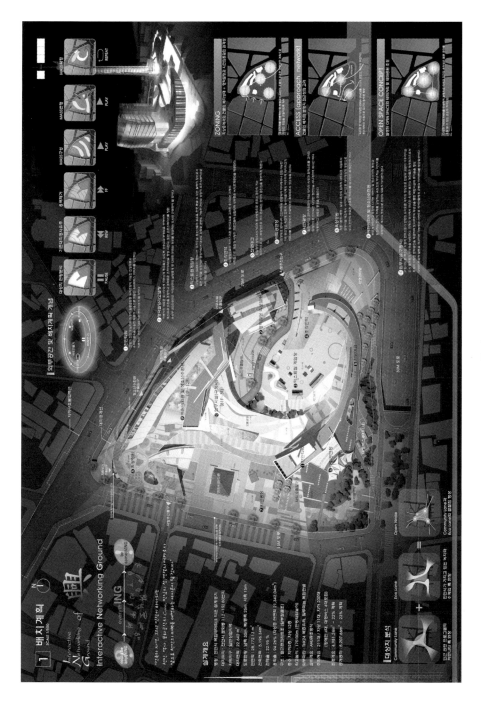

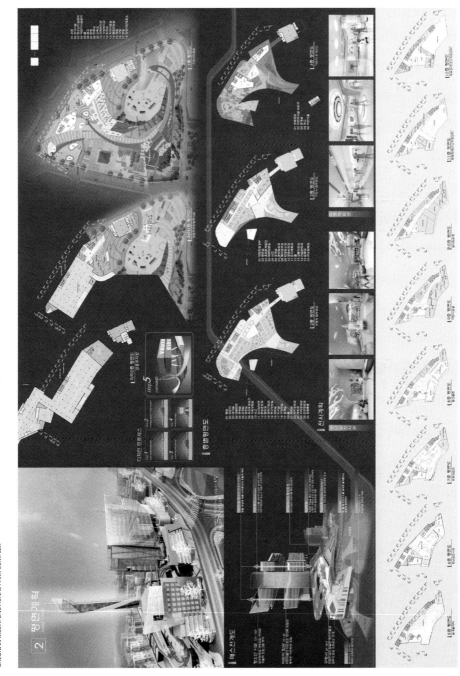

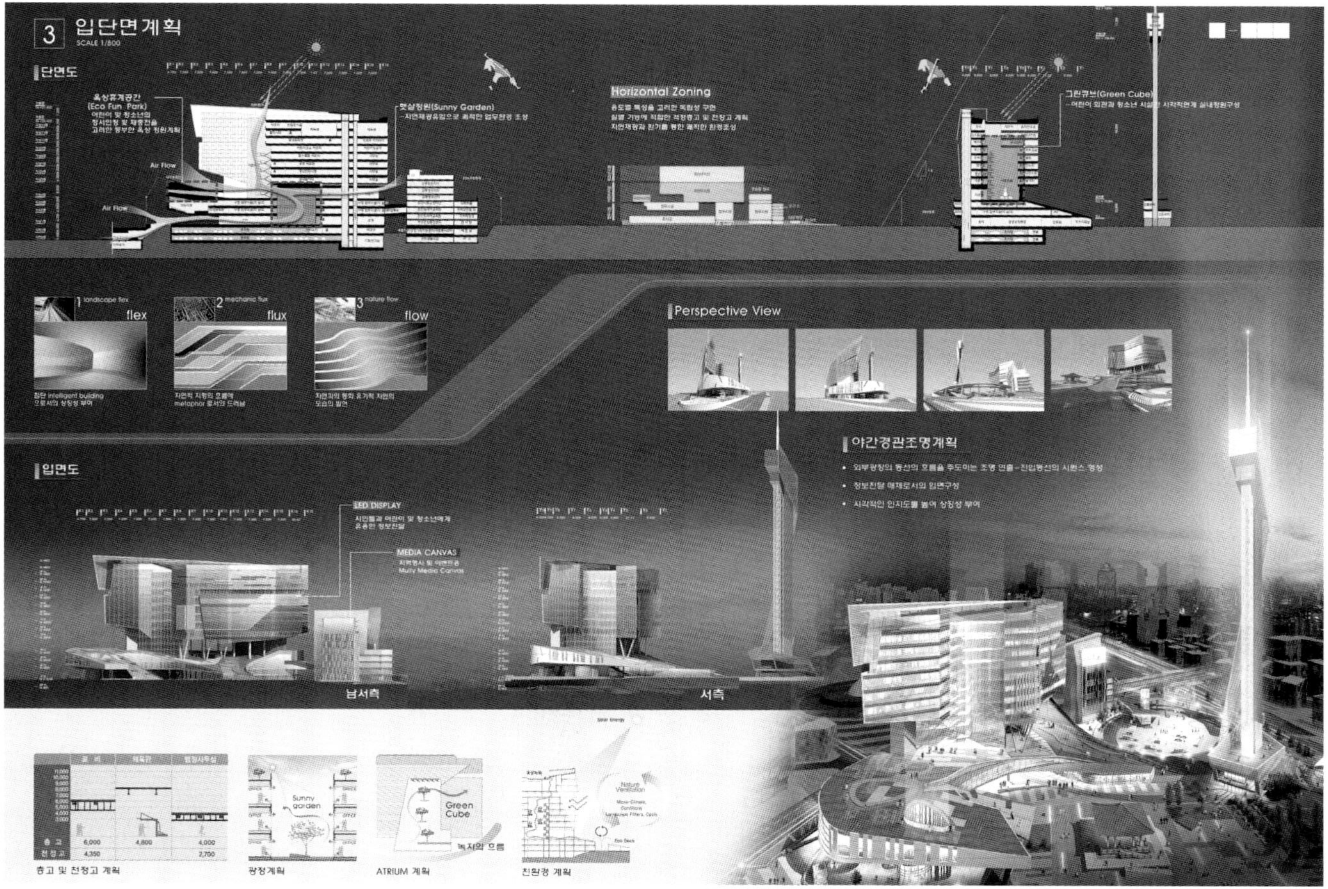
3 입단면계획
SCALE 1/800
단면도
옥상휴게공간
(Eco Fun Park)
어린이 및 청소년의
정서안정 및 재충전을
고려한 풍부한 옥상 정원계획
Air Flow
Air Flow
햇살정원(Sunny Garden)
-자연채광유입으로 쾌적한 업무환경 조성
Horizontal Zoning
용도별 특성을 고려한 독립성 구현
실별 기능에 적합한 적정층고 및 천정고 계획
자연채광과 환기를 통한 쾌적한 환경조성
그린큐브(Green Cube)
-어린이 회관과 청소년 시설간 시각적연계 실내정원구성
1 landscape flex
flex
첨단 intelligent building
으로서의 상징성 부여
2 mechanic flux
flux
자연적 지형의 흐름에
metaphor 로서의 드러냄
3 nature flow
flow
자연과의 동화 유기적 자연의
모습의 발현
Perspective View
야간경관조명계획
외부광장의 동선의 흐름을 주도하는 조명 연출-진입동선의 시퀀스 형성
정보전달 매체로서의 입면구성
시각적인 인지도를 높여 상징성 부여
입면도
LED DISPLAY
시민들과 어린이 및 청소년에게
유용한 정보전달
MEDIA CANVAS
지역행사 및 이벤트용
Multy Media Canvas
남서측
서측
층고
천정고
6,000
4,350
4,800
4,000
2,700
층고 및 천정고 계획
Sunny garden
광정계획
Green Cube
녹지의 흐름
ATRIUM 계획
Nature Ventilation
친환경 계획

\>> OFFICE

COMPLEX-THEME PARK TOWN OF CHEONAN

Location Munhwa-dong, Cheonan-si, Chungcheongnam-do, Korea Site Attribute General Commerce Site Area 22,642.00m² Building Area 8,550m² Total Floor Area 29,977m² Building Coverage Ratio 37.76% Floor Area Ratio 98.01% Structure RC, Steel, Truss Exterior Finish Titanium Steel Plate, T24 Pair Glass, Tempered Glass Parking 214 Cars Architecture Design Lee Geun-chang, Kim Yeong-chan | AUM & LEE Architects Associates + Jo Hyeong-sik | TOWOO Architects & Engineers Design Team AUM & LEE_Sim Jae-kun, Kim Yu-jin, Oh Seong-jin, Kim Jung-ho, Kim Dae-yeol, Du Gwang-ho, Kim Jong-min, Jeong Ji-yun, Lee Gye-yeong, Ryu Seung-yeon

SITUATION The site has a significant meaning in terms of location, history, topography and urbanism considering the features such as the geographical form of Oryongjaengju (five dragons fighting for cintamani), the heart of Cheonan's urban center, the triangle reminiscent of Cheonan Samgeori (Interchange), the declining old commercial districts, the starting point of old underground shopping malls and the street corner to Cheonan Station. However, numerous characterless low-rise buildings made the present conditions into an example of common monotonous city.
CONCEPT
Concept 1: Cultural Community - Urban Eye / Children's Center - Dream Village / Youth Training Center - Extreme Club / Public Office Facility - Info Zen / Park - Festival Park / Observatory - City View
As the core and eye of Cheonan, complex-theme park town will be a cultural destination of Cheonan that creates a new cultural community consisting of all generations according to the new family lifestyle.
Concept 2: ICON of Cheonan - Urban Tornado
The image of an ascending dragon was symbolized to bring changes and dynamism to the monotonous daily life of the city, establishing itself as the symbol of Cheonan, and at the same time the coiling tornado stretched from the ground to the sky became the motive of the organic from that unified the observatory and other facilities.
Concept 3: Cubical Space - Cubic Patio
- Soul_Rotating Observatory: Symbolism
- Body_Office Work, Youth Training Center: Floating, Unification
- Ground_Park, Children's Center: Publicity, Locating
- Root_Fiesta Valley Mall: Connecting Beodeul Yukgeori (Interchange) and the Underground shopping mall, Accessibility and Unification of Functions.
The observatory circulation starting from the central atrium zigzags and links various levels of floors, allowing the access to the observatory through the vertical atrium and offering a special experience to the users.
PLACE MAKING We put our priority on the city and citizens of Cheonan. Instead of inserting yet another building into a dense building forest, we floated it in the air and created a park below for the citizens, thus opening up the view and giving the space back to the citizens. By making spaces for leisure and cultural activities, the urban eye 'complex-theme park town of cheonan' is a new center of the city that will upgrade the level of daily life.

ATRIUM

PLAZA PLATE

RAINBOW BRIDGE

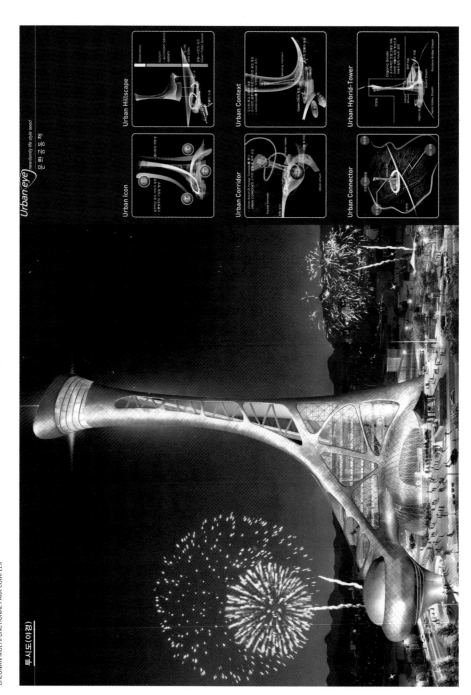

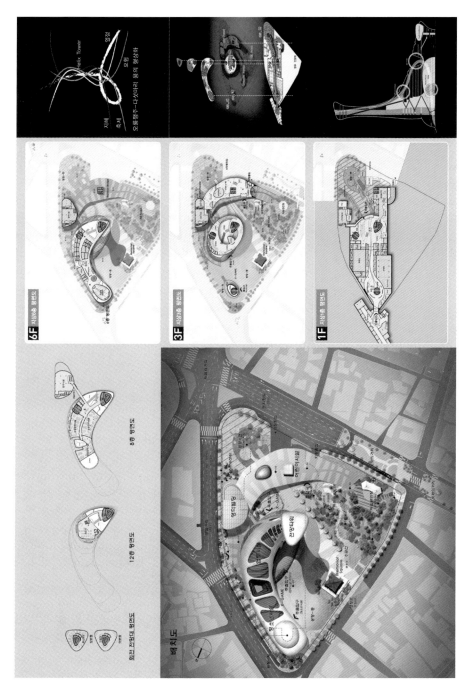

\>\> OFFICE

COMPLEX-THEME PARK TOWN OF CHEONAN

Location Munhwa-dong, Cheonan-si, Chungcheongnam-do, Korea **Site Attribute** General Commerce **Site Area** 22,642.00m² **Building Area** 6,401.17m² **Total Floor Area** 29,614.90m² **Building Coverage Ratio** 28.27% **Floor Area Ratio** 100.28% **Building Scope** B1~12F **Structure** Steel, RC **Exterior Finish** T24 Low-E Pair Glass, Exposed Concrete, High-Density Wood Panel, Metal Fabric **Interior Finish** Granite Rubbing, T3 Vinyl Tile, Eco-Paint, T15 Sound-Absorbing Tex **Parking** 216 Cars **Architecture Design** Kim Yong-gwon, Han Dae-jin, Cho Won-kyu | DOSHIIN Architecture + Lee Young-ik | DONGWOO Architects & Consultants + Kang Gye-suk | DS GROUP Architecture **Design Team** DOSHIIN_Choi Seop, Kim Seong-uk, Byeon Nam-il, Kim Su-jeong, Sin Bong-geun, Im Seung-hyeok, Lee Geun-taek / DONGWOO_Ryu Hui-sun, Lee Hye-kyeong, Kim Jeong-su, Yu Min-sang, Roh Chung-hwan

THE RISING CITY

The old urban system mixed with the trace of the past...
The insufficient public open spaces...
The decreasing floating population in Myeongdong...
We plug in the structure symbolizing a new force and the sense of direction.
The citizen's plaza, which reduces the stagnation of the surroundings, is linked to the streets in the vicinity. The newly transplanted paradigm is revitalized by the activities and gradual ripple effect to the site environment. The geographical identity of Cheonan, which has maintained the topography of 'Oryongjaengju, five dragons fighting for cintamani)', emits the image of the city moving toward the future.

DESIGN AIM
- To revitalize the old urban center and realize a complex-theme park town as a cultural center for Cheonan citizens.
- To realize architecture as a global landmark representing Cheonan: creating a symbolic space of environment-friendly and future-oriented Cheonan, providing outdoor spaces that accept various themes for the gatherings of Cheonan citizens, planning communities and open spaces that encourage the creative activities of young adults and children, promoting the economic revitalization of the Cheonan station square and the areas surrounding traditional markets, and planning a space for diverse cultural activities and events.

LAYOUT PLAN
- To actively accept various access levels to the center of the old urban network from all direction of the site.
- To plan the open square on the west and a group of buildings on the east to encircle the sunken square and the observatory.

FLOOR PLAN
- To enhance the spatial efficiency by overlapping multiple layers considering the urban structure that has reached the breaking point of the horizontal differentiation.
- To lift the ground using artificial boards in order to maximize the physical potential of the topography and connect the mass and the earth as one, connecting the indoor and outdoor spaces in a natural manner.

ELEVATION PLAN
- To introduce an urban device using the features of the site to emphasize the organic relation between architecture and the city and create a newly interpreted urban landscape.
- To give a symbolic flow to the existing urban spatial structure turned into slums and propose a direction of future developments on surrounding environments.

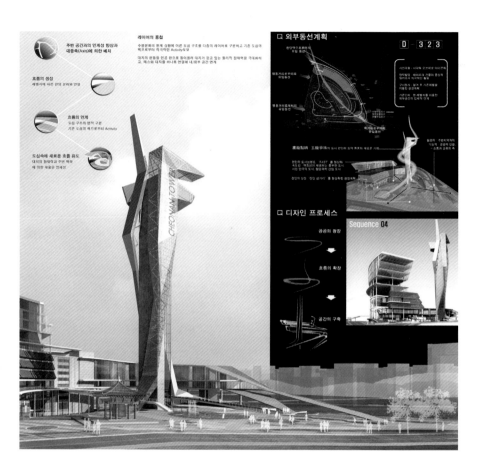

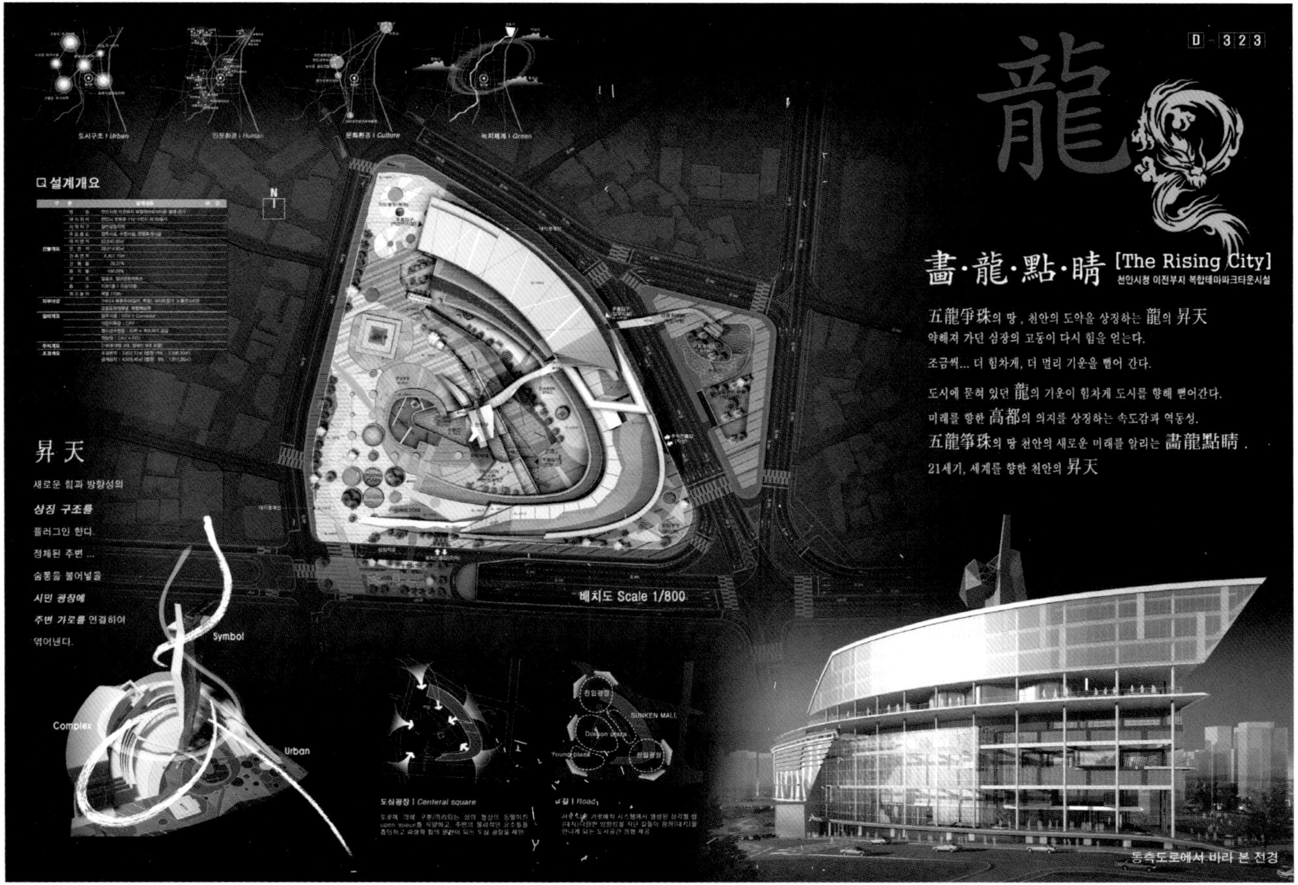
D - 3 2 3
龍
畵·龍·點·睛 [The Rising City]
천안시청 이전부지 복합테마파크타운시설
五龍爭珠의 땅 , 천안의 도약을 상징하는 龍의 昇天
약해져 가던 심장의 고동이 다시 힘을 얻는다.
조금씩... 더 힘차게, 더 멀리 기운을 뻗어 간다.
도시에 묻혀 있던 龍의 기운이 힘차게 도시를 향해 뻗어간다.
미래를 향한 高都의 의지를 상징하는 속도감과 역동성.
五龍爭珠의 땅 천안의 새로운 미래를 알리는 畵龍點睛 .
21세기, 세계를 향한 천안의 昇天
도시구조 I Urban
인문환경 I Human
문화환경 I Culture
녹지체계 I Green
□ 설계개요
N
昇 天
새로운 힘과 방향성의
상징 구조를
플러그인 한다.
정체된 주변 ...
숨통을 불어넣을
시민 광장에
주변 가로를 연결하여
엮어낸다.
Symbol
Complex
Urban
배치도 Scale 1/800
도심광장 I Centeral square
길 I Road
SUNKEN MALL
Dragon plaza
Young plaza
동측도로에서 바라 본 전경

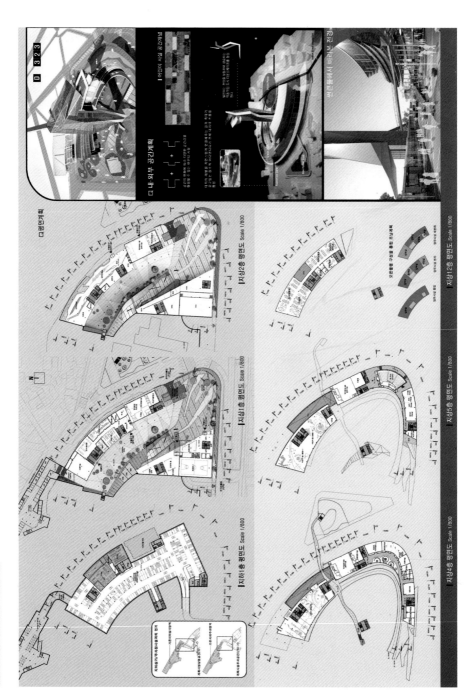